Meiju Award—Shortlisted Works of Asian-Pacific Interior Design
美居奖——亚太室内入围精选（下）

Sales Center & Club House 售楼处、会所空间　　**Hotel Space** 酒店空间　　**Exhibition Space** 展览空间

佳图文化　编

图书在版编目（CIP）数据

美居奖：亚太室内入围精选 . 下 / 佳图文化编 . —
天津：天津大学出版社，2014.5
ISBN 978-7-5618-5081-7

Ⅰ.①美… Ⅱ.①佳… Ⅲ.①室内装饰设计—作品集—中国—现代 Ⅳ.① TU238

中国版本图书馆 CIP 数据核字 (2014) 第 112738 号

出版发行	天津大学出版社
出 版 人	杨 欢
地　　址	天津市卫津路 92 号天津大学内（邮编：300072）
电　　话	发行部：022—27403647
网　　址	publish.tju.edu.cn
印　　刷	利丰雅高印刷（深圳）有限公司
经　　销	全国各地新华书店
开　　本	245mm×325mm
印　　张	18
字　　数	351 千
版　　次	2014 年 6 月第 1 版
印　　次	2014 年 6 月第 1 次
定　　价	320.00 元

凡购本书，如有质量问题，请向我社发行部门联系调换

Preface 前言

Initiated by JTart Publishing & Media Group and sponsored by the real estate design industry's famous and professional *New House* magazine, "Meiju Award" is a comprehensive award on real estate design and the most authoritative award in this field. Based on the principle of fairness and openness, JTart Publishing & Media Group invited numerous famous designers, design firms and expert teams to screen the shortlisted projects from a professional and high-standard angle, to make sure that every case is conformed to the evaluation criteria and spiritual connotation of "Meiju Award", thus achieving the award's greatest significance.

This set of books collects some of the shortlisted interior design works, covering office space, commercial space, exhibition space, restaurant space, club space, hotel space and other public spaces. The cases are analyzed from several perspectives, such as design style, design philosophy, design feature and material selection, and each of them is illustrated with detailed technical drawings and realistic pictures, which present the design details and design points in an all-round way. With concise description, abundant pictures, professional typesetting and exquisite bookbinding, this set of books aims to provide a learning and communication platform for excellent designers and amateurs.

"美居奖"是由佳图出版传媒集团发起,业内著名地产设计专业杂志《新楼盘》杂志社主办的房地产设计类的综合性大奖,亦是地产设计行业内最权威的专业奖项。本着"公平、公正、公开"的评选原则,佳图出版传媒集团邀请众多著名的设计师、设计公司以及专家团队从专业的、高标准的角度筛选入围项目,力求使每一个案例都符合"美居奖"的评选标准与精神内涵,从而实现"美居奖"设置的重大意义。

本套书辑录了入围"美居奖"的部分室内设计项目,内容涵盖办公空间、商业空间、展览空间、餐饮空间、会所空间、酒店空间以及其他公共空间的室内设计。案例分析主要从室内设计风格、设计理念、设计特色、材料选择等方面入手,并且每个案例均配备了翔实的设计技术图纸以及实景图,全方位地展示设计细节和设计要点。本套书设计说明文字简明扼要、图片丰富、排版专业、装帧精美,旨在为优秀的设计师及相关人士提供一个学习交流的平台。

CONTENTS

Sales Center & Club House 售楼处、会所空间

008 Sales Center of Vanke Sun Moon Park, Dalian
大连万科朗润园销售中心

014 Sales Center of Vanke Sakura Garden, Dalian
大连万科樱花园销售中心

020 T-Share International Office Tower, Shenzhen
深圳田厦国际中心

026 Fusion Club
融会会所

036 Taimen Wine Club
台门酒会

046 Butterfly Therapeutic Retreat
意兰庭保健会所

058 Wuxi Lingshan Vihara
无锡灵山精舍

068 Understand World Tea
观茶天下茶室

086 Yincheng Red Wine Club
银城红酒会所

096 Wujiang Shengze Jiacheng International Hotel Nightclub
吴江盛泽嘉诚国际酒店夜总会

108 Fantasy KTV
奇幻量贩式KTV

Hotel Space 酒店空间

116 Beijing Shouzhou Grand Hotel
北京寿州大饭店

136 Jiangsu Tianmu Brilliant SPA Holiday Hotel
江苏天目辉煌温泉度假酒店

144 Radisson Blu Hotel Liuzhou
柳州丽笙酒店

目录

156 Pan Pacific Ningbo Hotel
     宁波泛太平洋大酒店

164 Xianghe Bainian Hotel
     祥和百年酒店

180 Chongqing Weisilai Xiyue Hotel
     重庆威斯莱喜悦酒店

190 Juchunyuan Boutique Hotel
     聚春园驿馆

200 Radegast Lake View Hotel, Beijing
     北京康源瑞廷酒店

208 Jintai Longyue Seaview Golf Resort, Liaoning
     辽宁金泰珑悦海景高尔夫度假酒店

Exhibition Space 展览空间

218 Huizhou Harmony World Watch Exhibition Hall
     惠州亨吉利世界名表展厅

226 Kunshan Xupin Exhibition Hall
     昆山叙品展厅

234 Phase II Exhibition Hall of Shenzhen O'seka Art Exhibition Center
     深圳奥斯卡艺展中心二期展厅

246 Hangzhou A·Base Furnishing & Salon
     杭州佰色 A·Base 陈设 & 沙龙

260 Wuhan Zhongda Jiangbao 4S Store Exhibition Hall
     武汉中达江宝 4S 店展厅

266 Restonic Furniture Exhibition Stand
     运时通家具展位

272 Three Forks Lake Digital Display Space
     三岔湖数字展示空间

278 Original Fengjing Public Facilities
     原风景大楼公共设施

Sales Center & Club House
售楼处、会所空间

- Theme and Style 主题风格
- Cultural Connotation 文化意蕴
- Atmosphere Creation 氛围营造
- Material Texture 材质肌理

KEYWORDS 关键词

NATURAL SPACE
自然空间

HOLLOW MODEL
镂空造型

GREEN & ECOLOGICAL
绿色生态

Furnishings / Materials
软装 / 材料

BLACK FOREST MARBLE
古木纹大理石

WHITE ARTIFICIAL STONE
白色人造石

GLACIER GRAY MARBLE
冰川灰大理石

SILVER MIRROR 银镜

BLACK MIRROR STAINLESS STEEL 黑色镜面不锈钢

PERFORATED ALUMINUM PANEL 穿孔铝板

OAK VENEER
橡木饰面

Sales Center of Vanke Sun Moon Park, Dalian

大连万科朗润园销售中心

Designer: Yu Qiang
Design Company: YuQiang & Partners Interior Design
Location: Dalian, Liaoning, China
Area: 600 m²

设 计 师：于强
设计公司：于强室内设计师事务所
项目地点：中国辽宁省大连市
面　　积：600 m²

A spatial setting looking out to the courtyard with flickering tree is created by introducing such outdoor natural elements as log, green color and so forth into the interior, and by the use of the geometric cut-out patterns and soft light. The pleasant surprise brought by this placid contentment engenders a feeling of authenticity and lavish abundance with intimacy to Mother Nature. This brings about a natural spatial realm of uncluttered simplicity, comfort, and all-season greenery for the northern city.

通过原木、绿色等元素将户外的树木等自然景观延伸至室内，通过运用镂空的几何图案造型、柔和的灯光，营造一个面向庭院、树影摇曳的空间，以此带来平淡的惊喜，感受一种亲近自然的真实和奢华，给这个北方城市带来一个简单、舒适、四季常青的自然空间。

First Floor Plan
一层平面图

KEYWORDS 关键词

FLEXIBLE SPACE
灵动空间

NATURAL & ECOLOGICAL
自然生态

HARMONIOUS ENVIRONMENT
环境和谐

Furnishings / Materials
软装 / 材料

SNOW WHITE MARBLE
雪花白大理石

WHITE ALUMINUM PANEL
白色铝板

PERVIOUS MARBLE
透光云石

OAK LOGS
橡木原木

TEAK
泰柚

BOLON CARPET
Bolon地毯

Sales Center of Vanke Sakura Garden, Dalian
大连万科樱花园销售中心

Designer: Yu Qiang
Design Company: YuQiang & Partners Interior Design
Location: Dalian, Liaoning, China
Area: 739 m²

设 计 师：于强
设计公司：于强室内设计师事务所
项目地点：中国辽宁省大连市
面　　积：739 m²

The conceptualization here evolves from the cherry blossom. The handling of the space has succeeded in setting free from the dwelling's inherent "boxy" appearance by adopting a set of broken lines to traverse the space and produce a decomposition of the spatial expanse, as the geometric forms and zigzag interfaces form an inspirational response to the enchanting sensation of the surrounding environment's embrace of lush green hills. A color palette extends and enhances the elegant white and pastel pink of cherry blossom from the window, with a set of snow-white striated patterns, hardwood outlines, and light grey leather covering to be complemented by matching natural wooden chairs to embody an eco concept, ensuring that the spatial ambience is even more intimately integrated with Mother Nature.

本案以樱花为元素，展开构思。空间上，打破原建筑固有的"盒子"形体，采用折线来穿插、分解空间，抽象的几何形体、转折起伏的界面与环境中叠山环绕的灵动感形成呼应。色彩延续窗外樱花高雅的白色与粉色、细纹雪花白图案、实木线条、浅灰色皮革配以原木座椅，体现生态理念，使整个空间氛围更加贴近自然。

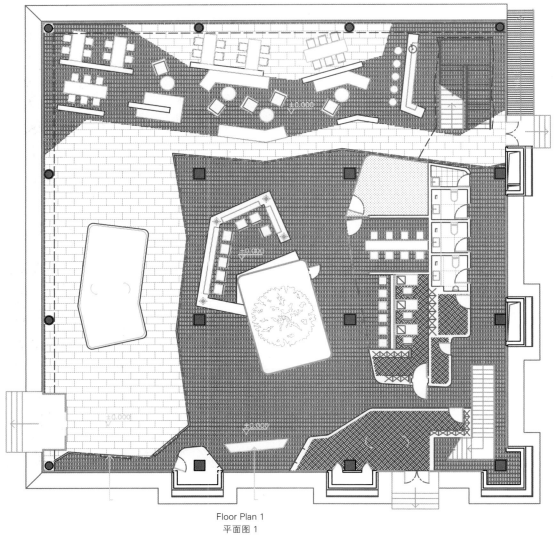

Floor Plan 1
平面图 1

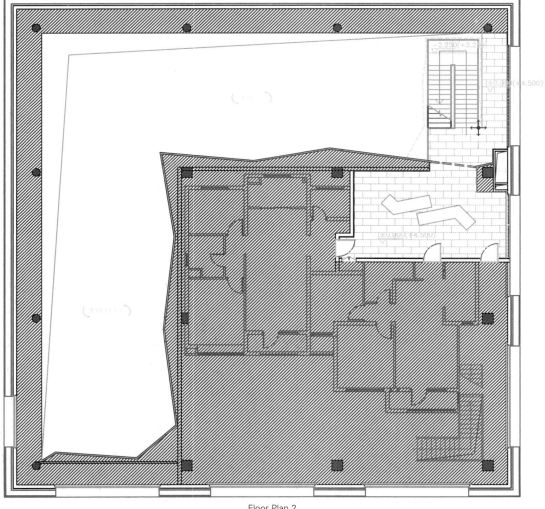

Floor Plan 2
平面图 2

KEYWORDS 关键词

GEOMETRICAL SHAPE
几何造型

LINEAR ELEMENT
线条元素

FASHIONABLE STYLE
时尚气息

Furnishings / Materials
软装 / 材料

WHITE MARBLE
直纹白大理石

PARIS GRAY MARBLE
巴黎灰大理石

BLACK MIRROR STEEL
黑镜钢

WALNUT WOOD FLOOR
胡桃木地板

WOVEN CARPET
编织地毯

GRAY OAK VENEER
灰橡木饰面

T-Share International Office Tower, Shenzhen

深圳田厦国际中心

Designer: Yu Qiang
Design Company: YuQiang & Partners Interior Design
Location: Shenzhen, Guangdong, China
Area: 1,000 m²

设 计 师：于强
设计公司：于强室内设计师事务所
项目地点：中国广东省深圳市
面　　积：1 000 m²

The suspended decoration with geometric pentagonal shapes and a mass of fine lines is extending to fill up the whole space. The floor is paved with black-and-white wood grain marble. The application of further linear elements makes the interior space retain the inherent rational quality of the building whilst imparting a certain subtle ambiance to it. A series of geometric cut-out patterns suspended from the ceiling divides the space into two distinct areas—negotiation area and exhibition area, open but private. The embellishment of metal material brings some fashionable elements to this space.

带有五边形几何镂空图案的悬吊造型充满了整个空间，密密的细线叠加在造型上，地面采用黑白木纹大理石，直线元素的运用使得室内空间既保留了建筑自身理性的特点，又增添了些许柔美的气息。几何图案的镂空造型作为分隔空间的装置，把空间分为洽谈及展示两个大的区域，既开敞又不失私密感。金属材质的点缀，增加了空间的时尚感。

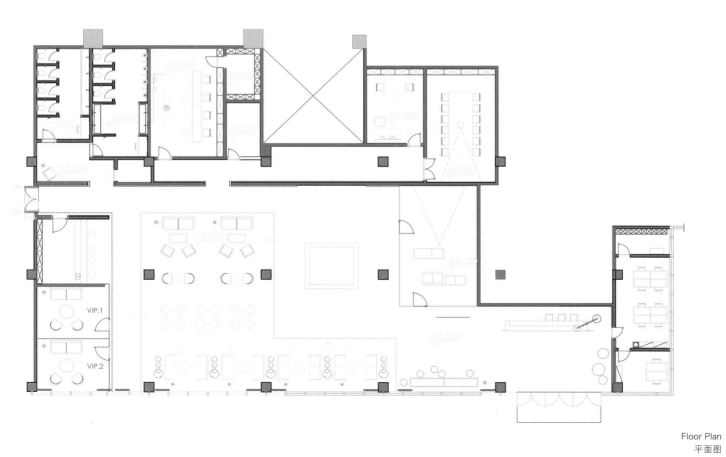

Floor Plan
平面图

KEYWORDS 关键词

TASTEFUL SPACE
格调空间

STEADY & GRACEFUL
沉稳大气

COLOR LOGIC
色彩逻辑

Furnishings / Materials
软装 / 材料

OLD WOODEN BOARD
老木板

FRAXINUS MANDSHURICA STAINING 水曲柳染色

BLACK TRAVERTINE
黑洞石

COPPER WIRE DRAWING
拉丝铜

HORSE FUR
马毛皮

WOOD-FIBER ACOUSTIC PANEL 木丝吸音板

Fusion Club
融会会所

Chief Designer: Feng Jiayun
Participating Designer: Tie Zhu, Geng Shunfeng
Design Company: Wuxi S-zona Designer Consultant
Location: Wuxi, Jiangsu, China
Area: 1,440 m²

主设计师：冯嘉云
参与设计师：铁柱、耿顺峰
设计公司：无锡市上瑞元筑设计制作有限公司
项目地点：中国江苏省无锡市
面　　积：1 440 m²

The mottled, ancient and whirling space texture and the vivid and massive historical memory are identical with the glorious and brilliant "Chinese modern industry and commerce" and "Republic of China". Shaping story is the original intention of the design, and at the same time, intellectual and tasteful space is also the correspondent expectation based on the high-end target customers' psychological mechanism. The property of club determines that the club is doomed to be a body and mind relaxing place for a small group, and an exclusive field for the city upstarts' "late luxury, slow life". Therefore, the color tone adopts the international performance of grey, and the integral whole, steady and graceful attributes suggest the care of the noble spirit. The black leather, grey-blue wallpaper, cloth art, gray stone, and the magnificent and natural wood grain, camel carpet, brown chair back, desk sets and deep yellow cowhide, all show the color logic of natural transition from cold to warm tone; and the abundant contrasts of materials and the change of decorative pattern form vivid space tension, which is restraining but active.

斑驳、古意、婆娑的空间肌理，带有鲜明、厚重的历史记忆，与曾经辉煌的"中国近现代工商业"、"民国"在气质上相吻合。塑造故事，成为设计的初衷，同时，知性、雅致的空间，亦建立在与高端目标客群心理机制相对应的预期之上。会所业态，注定是一小族群的身心归所，是城市新贵"后奢侈、慢生活"的专属场所。为此，在色彩基调上，采用国际化表现手法的灰调，以浑然一体、沉稳大气暗示对贵族精神的关注。从黑的皮革、灰蓝的墙纸、布艺及灰色水纹的石材，到瑰丽大方的木纹、驼色的地毯、褐色的椅背及桌套、深黄的牛皮，无不演绎着由冷色调到暖色调的自然过渡与缜密的色彩逻辑，并由丰富的材质对比、纹饰变化形成了生动的空间张力，内敛中洋溢着悦动。

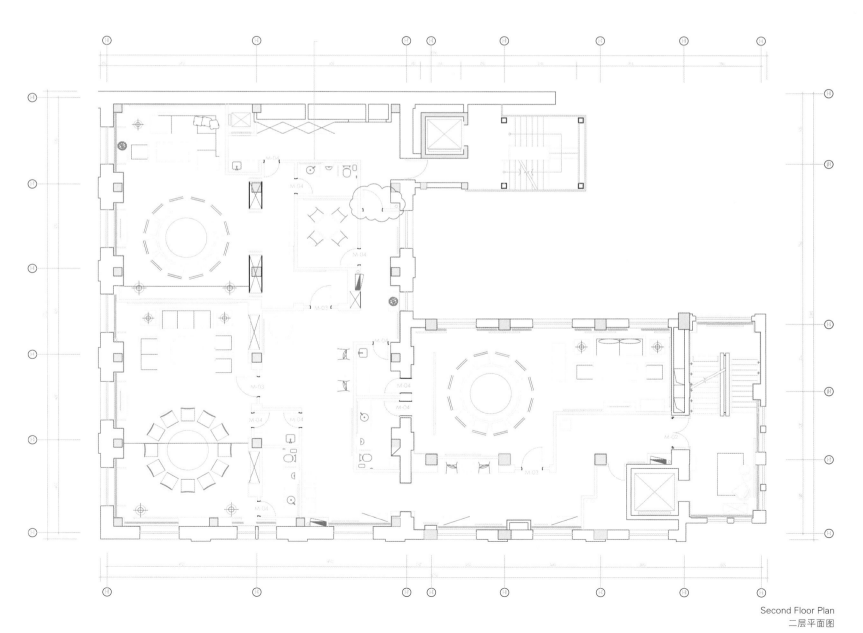

Second Floor Plan
二层平面图

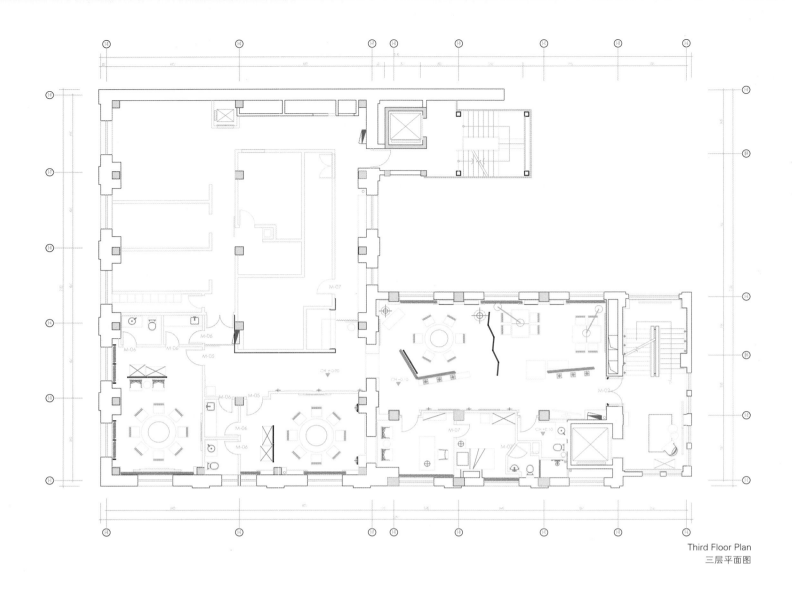

Third Floor Plan
三层平面图

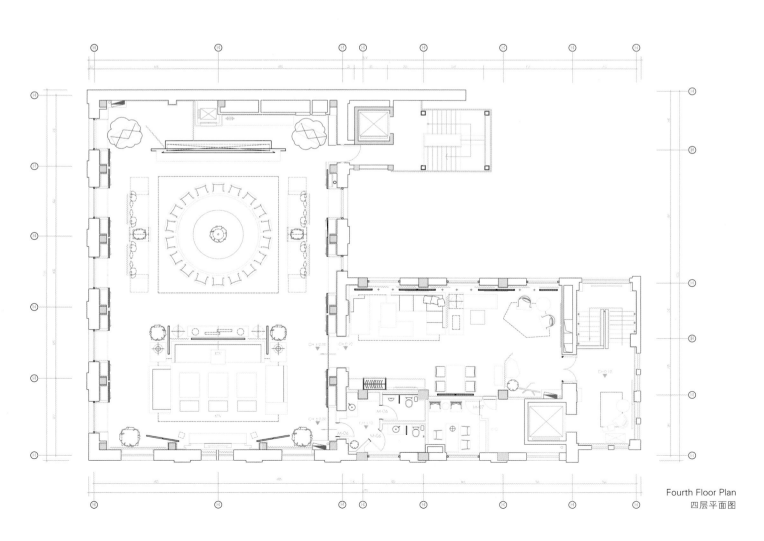

Fourth Floor Plan
四层平面图

KEYWORDS 关键词

NEW CHINESE STYLE
新中式

UNSOPHISTICATED & PEACEFUL 古朴静谧

CULTURE APPEAL
文化韵味

Furnishings / Materials
软装 / 材料

BLUESTONE
青石

DEAL BOARD
松木板

ANTIQUE SOLID WOOD FLOOR 仿古实木地板

PARCHMENT LAMP
羊皮纸灯

Taimen Wine Club
台门酒会

Designer: Lin Sen, Xie Guoxing
Design Company: Hangzhou Gazer Decoration Design
Location: Hangzhou, Zhejiang, China
Area: 360 m²

设 计 师：林森、谢国兴
设计公司：杭州肯思装饰设计事务所
项目地点：中国浙江省杭州市
面   积：360 m²

The project is located at the foot of the Chenghuang Mountain in Hangzhou, close to the Hefang Street and Southern Song Imperial Street. Hefang Street is an ancient street with a long history and deep cultural deposits. It used to be "the root of the imperial palace" in the ancient capital city of Hangzhou, and also the cultural and economic center of the Southern Song Dynasty.

White walls and black tiles can be seen from the front door. When entering into the room, you can see antique solid wood floor, elegant column and the wooden partition wall, full of "antique flavor". Although their colors are not so striking, the collocation is exciting. What a joy of such a kind of quiet and primitive simplicity it is!

The structure of the project is divided into two floors. The first floor includes the tasting area, sales area, cultural exhibition, reception area and operation room mainly for the individual guests; while the second floor is mainly the VIP tasting activity area. The designers put the essence —"dated, new, luxury, pure" — of the old Taimen throughout the design.

The designers mainly adopt the technique of new Chinese style on the first floor, and keep the gentle cultural appeal of Shaoxing wine, blending the wine, calligraphy and painting and natural atmosphere. The original building is a wooden house with quite a lot of pillars, which is the characteristic which the later idea of "pillar forest" is based on. The later added pillars serve as the support of the sun roof, and play an important role in regional coherence and culture presentation as well. The winding stream passes through the pillars, which reminds people of the scene of ancients' drinking and writing poems, and also increases the interest of wine tasting. On the second floor, white parchment lamps hanging on the ceiling have peculiar modeling, which is the greatest characteristic of this case. The white paper lamps hide the line pipes on the top and retain the original high space as well, coordinating the atmosphere and color tone of the whole space. The second floor has three bays and can be divided and closed through the white gauze; and the white gauze is used to decorate the space, making it pure white and flawless, quiet and gentle.

　　本案位于杭州城皇山脚下，河坊街与南宋御街旁。河坊街是一条有着悠久历史和深厚文化底蕴的古街，它曾是古代都城杭州的"皇城根儿"，更是南宋的文化中心和经贸中心。

　　白墙、黑瓦，从门面开始便很清新。进入房间，是一室的"古色古香"，仿古实木地板、大气的柱子以及用木材做的背景隔断，用材用色虽不出奇，但搭配得让人很欣喜。这样的安静与古朴，令人何等欢喜！

　　本案在结构上分为上、下两层，一层主要为散客参观品尝区、售卖区、文化展示区、总台区以及操作间；二层则主要是VIP品鉴活动区。在设计上，设计师将老台门的精髓"陈、新、奢、纯"贯穿始终。

　　一楼主要采用新中式的手法，保留了绍兴老酒温婉的文化韵味，将酒、字画及自然的氛围融入其中。原建筑为木结构的房子，柱子较多，正是依据这个特点，才有了后来柱林的创意。后来加入的柱子既作为阳光屋顶的支撑，又起到了区域连贯和文化展示的作用；蜿蜒的小溪横穿其中，让人联想到古人饮酒赋诗的场景，亦增加了品酒的情趣。上到二楼，楼梯顶面排列的白色羊皮纸灯，造型奇特，当属本案的最大特点。白色的纸灯既遮挡了顶部穿插的线管，又保留了原有的高挑空间，协调了整个空间的氛围、色调。二楼为三开间，通过白纱帘，既可分又可合，而白纱帘在这里更多的是装饰整个空间，使空间洁白无瑕，静谧温婉。

三伏 采药

陈年 窖藏

KEYWORDS 关键词

HUI-STYLE ELEMENT
徽派元素

POETIC SPACE
诗意空间

COMFORTABLE & QUIET
舒适静谧

Furnishings / Materials
软装 / 材料

ANCIENT WOOD GRAIN PANELS　古木纹饰面板

SMALL BLUE BRICKS
小青砖

SESAME BLACK STONE
芝麻黑石材

ANTIQUE PLATE
仿古板

TILES
瓦片

CURTAIN
幔帐

Butterfly Therapeutic Retreat
意兰庭保健会所

Chief Designer: Xu Jianguo
Participating Designer: Chen Tao, Ouyang Kun, Cheng Yingya
Design Company: Hefei Xu Jianguo Architecture & Interior Design Co., Ltd.
Location: Hefei, Anhui, China
Area: 460 m²
Photography: Wu Hui

主设计师：许建国
参与设计师：陈涛、欧阳坤、程迎亚
设计公司：合肥许建国建筑室内装饰设计有限公司
项目地点：中国安徽省合肥市
面　　积：460 m²
摄　　影：吴辉

In order to seek a state of mind, place a kind of emotion, and also create a comfortable, natural, quiet, and relaxing space, and the mental space the public expect for in the noisy city, the designers use the artistic conception of the poem "By Chance" to present this project. The designers apply the Hui-style elements to the design, creating a reasonable space. The rare pure and fresh Chinese architectural style contains abundant Zen. The use of tiles is like the ink painting, and a few stacks of materials make people find everything new and fresh. Especially the use of the curtain softens the whole space. The design of the overall interior space is elegant, quiet, and full of poetic flavor and interest.

设计师借《偶然》这首诗的意境来表达本案，显然是为了寻求一种心境，寄托一种情感，亦是为了在喧嚣的闹市中，打造一个舒适自然、安静放松的空间，创造一个大众所寻觅的心灵空间。设计师将徽派元素融入设计中，整合出最合理的空间。少见的、清新的中式风格，蕴含禅意。瓦片的运用，就像水墨画一样；材料的少量堆砌，让人耳目一新。特别是幔帐的运用，柔化了整个空间。整个室内空间的设计幽雅、安静，富有诗意与情趣。

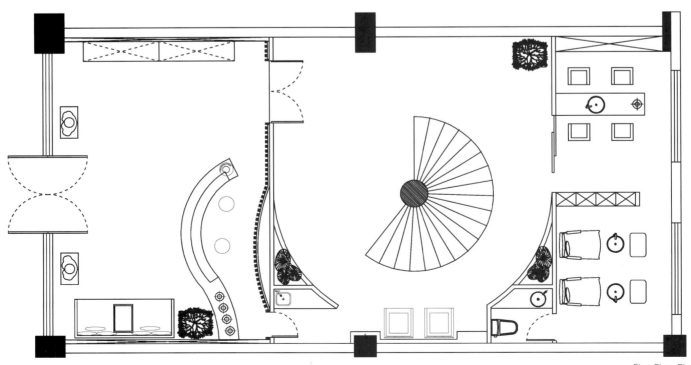

First Floor Plan
一层平面图

047

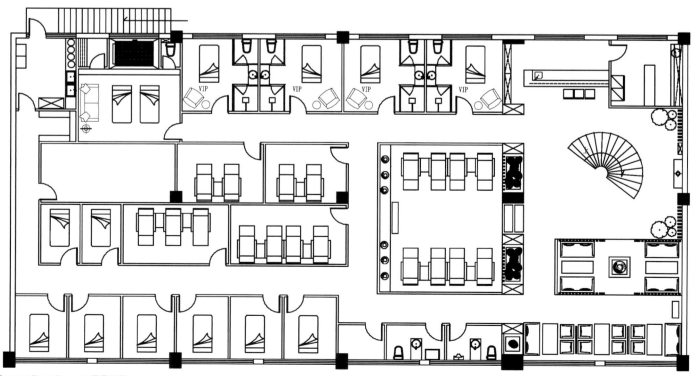

Second Floor Plan　二层平面图

KEYWORDS 关键词

ZEN SPACE
禅意空间

SIMPLE & PLAIN
简约质朴

EXQUISITE & ELEGANT
精巧素雅

Furnishings / Materials
软装 / 材料

WOODEN GRILLE
木质格栅

PENDANT LAMPS
吊灯

BAMBOO CURTAIN
竹帘

FURNITURE
家具

Wuxi Lingshan Vihara
无锡灵山精舍

Chief Designer: Lu Rong
Participating Designer: Shen Xi, Tian Jun, Li Ting
Design Company: HKG Group
Location: Wuxi, Jiangsu, China
Area: 9,800 m²

主设计师：陆嵘
参与设计师：慎曦、田珺、李婷
设计公司：HKG Group
项目地点：中国江苏省无锡市
面　　积：9 800 m²

The project is located in Lingshan, Wuxi, next to Lingshan Buddha with an area of approx. 9,800 m². 90 guestrooms are nestled in a tranquil bamboo forest. Clients could experience Zen ambiance here as well as the related activities held by Wuxi Lingshan Vihara.

As for the interior design of the vihara, the designers take bamboo as the theme to carry on the ambiance of bamboo vihara in India thousand years ago. In the lobby, the wooden grilles with sense of weathering, reception desk made of old copper, and the great bamboo pendant lamps make guests feel peaceful. In the simple but exquisite guestrooms, people could enjoy the small yard through the bamboo curtains. The furniture in tearoom is simple but full of Zen, available for the guests' regulation of mental state during tea time. Wuxi Lingshan Vihara with Zen theme provides the tuition, enlightenment and conception of Zen.

无锡灵山精舍坐落于无锡灵山胜境内，毗邻灵山大佛，总建筑面积9 800 m²，拥有约90间客房，掩映在一片安静的竹林之中。在这里，客人可以静下心来修身养性、体悟禅境，并参加精舍提供的各项与参禅相关的活动。

在精舍的室内设计中，设计师以竹为主题，传承了佛陀千年前在印度竹林精舍时的意境。进入大堂，颇具风化感的木制条形格栅天顶、旧铜打造的前台、几盏竹制的大吊灯，让人的心一下子沉静下来。朴素的客房，简单却很精巧，透过细密的竹帘，目光可以穿越到窗后富有禅意的小院子里。茶室里的家具简约却透着禅境，让客人在参茶的过程中更好地调节心境。无锡灵山精舍正是以"禅"为主题，提供给客人"禅"的教诲、"禅"的感悟、"禅"的意境。

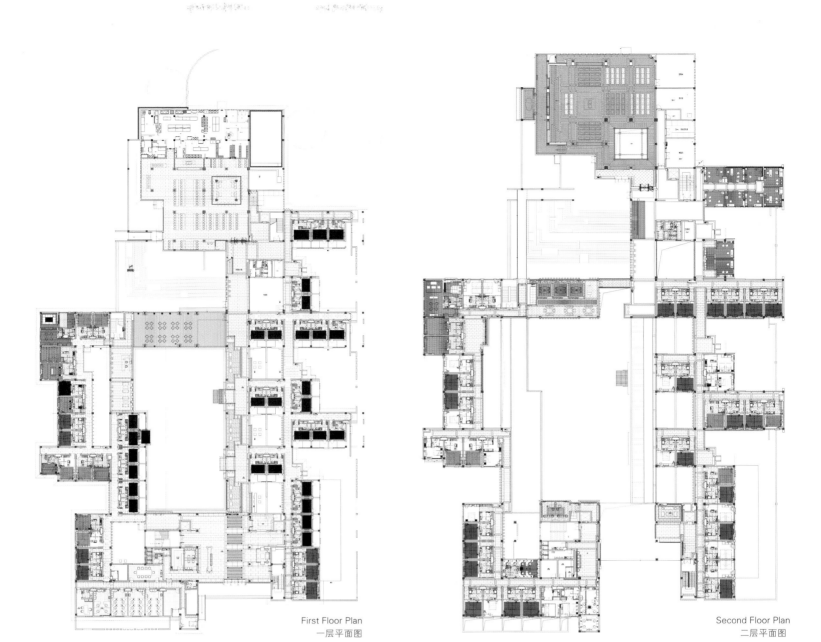

First Floor Plan
一层平面图

Second Floor Plan
二层平面图

KEYWORDS 关键词

HUI-STYLE
徽派风格

CULTURAL SPACE
文化空间

EASY & GRACEFUL
轻松优雅

Furnishings / Materials
软装 / 材料

ANCIENT WOOD GRAIN PANEL 古木纹饰面板

SMALL BLUE BRICK
小青砖

SESAME BLACK STONE
芝麻黑石材

ANTIQUE PLATE
仿古板

Understand World Tea
观茶天下茶室

Chief Designer: Xu Jianguo
Participating Designer: Chen Tao, Ouyang Kun, Cheng Yingya
Design Company: Hefei Xu Jianguo Architecture & Interior Design Co., Ltd.
Location: Hefei, Anhui, China
Area: 360 m²
Photography: Wu Hui

主设计师：许建国
参与设计师：陈涛、欧阳坤、程迎亚
设计公司：合肥许建国建筑室内装饰设计有限公司
项目地点：中国安徽省合肥市
面　　积：360 m²
摄　　影：吴辉

The project is located in the Mid-ring Town, Huangshan Road (the original Xuefu Road) in Hefei, which is the main culture street. Designers choose the Hui-style of strong tea culture background to reveal the characteristics of the project and try to create an idyllic and beautiful area and break the traditional Hui-style architectural features, so as to let people enjoy a relaxing and elegant environment, and experience the Huizhou tea culture essence. The orderly arranged horsehead walls are used in this case to enhance the Hui culture, which can draw people's attention to the natural and pure land.

The design idea focuses on Huizhou tea culture essence. The so-called "good wine can attract many guests and good tea can serve friends from afar" is exactly what the designers want to convey. The essence of Huizhou Tea is to establish virtue by tea, influence one's sentiment by tea, make friends by tea and entertain the guests by tea. The key of the design work is to create an environment and atmosphere of teahouse, thus really expressing the extensive and profound Chinese tea culture.

The first floor is the tea sales area, and the second floor is the tea tasting area. There is a shelflike partition wall in the entrance hall to reduce the influence of external environment to the inside. The first floor is divided into reception area, experience area, leisure landscape area and tea exhibition area. There are some wells to be parted in the middle of the tea exhibition area where some tea products are orderly displayed. The circular channels around the exhibition area are convenient for moving and selection. There is Guqin, book shelves, Guanyin figure, rockery waterscape in the leisure landscape area on the first floor, which makes people feel quiet, simple, gentle and natural. The designers apply artificial patio to this case. The rockery waterscape connects the first and second floor subtly. People can view the artificial patio on the first floor, extremely transparent, and the daylighting is also very good. Customers can enjoy the setting of the first floor by moving around the patio on the second floor. The landscape of the crane and water forms a delightful contrast, giving people a feeling of returning to the nature. Tea tasting area on the second floor is multifunctional, consisting of the service area, leisure area, calligraphy and painting area, lounge area, which meet the requirements of different customers. There is also a refrigerated tea store area which is convenient to store tea for the customers and entertain the guests. The feature of Hui-style architecture is that the rain in every direction on the roof should fall together in the patio. In view of the patio above and waterscape below, the designers are inclined to interpenetrate

观茶天下
UNDERSTAND ORLD T A
观茶天下
UNDERSTAN D WORLD TEA
大陆区域加盟咨询电话：400-616-1096
观茶天下

the indoor landscape of the first and second floor mutually, which makes each area seem independent, but merge into a whole.

The style of the project is explicit and the concept of the teahouse is different from the past ones. It refines the classic features of the traditional decorative elements to evolve new design symbols, using delicate and cultural decoration in the tea tasting area on the second floor. With the contemporary and concise design language, such a tea culture space is closer to the modern life. The liquidity, transparency, openness and integration of the space fully embody the art concept of the designers and the whole space: even in the complex metropolis, people are still able to create a pure sky.

本案位于合肥市黄山路(原学府路)中环城,这里是文化一脉相承的主街。设计师选择具有浓厚茶文化底蕴的徽派风格来彰显本案特点,创造一个世外桃源,试图打破传统徽派建筑的特点,让人享受一个放松、优雅的环境,细细体会徽州茶文化精髓。本案外观运用马头墙有序排列,可以增强徽州文化印象,让人容易注意到这片自然的净土。

设计思路主抓徽州茶文化精髓,"酒好能引八方客,茶香可会千里友"正是设计师所要表达的实质。徽州茶道,讲究以茶立德,以茶陶情,以茶会友,以茶敬宾。

设计工作的重点是营造茶楼环境、气氛，以求汤清、气清、心清；境雅、器雅、人雅，真正表达博大精深的中华茶文化。

本案一楼是茶叶销售区，二楼是品茶区。门厅处运用书架式隔断，减少外部环境对内部的影响。一楼分为前厅接待区、体验区、休闲景观区、茶叶展示区。茶叶展示区中间有水井相隔，展区有序地摆放着茶产品，四周有循环通道，方便流动与选取。休闲景观区有古琴、书卷架、观音、假山水景，让人感受平静、朴素、平和、自然的空间氛围。设计师把人造天井运用在本案中，其间的假山水景，巧妙地连接一、二层，在一楼可以看到人造天井，异常通透，采光效果好；二楼顾客可以围绕天井欣赏一楼布景，鹤与流水的造景相映成趣，给人一种回归自然与纯朴的感觉。二楼品茶区分服务区、休闲区、书画区、卧榻区，功能齐全，以满足不同客人的需求。另外还设有冷藏储茶区，将客人所购买的茶叶储藏，方便顾客待客之需。徽派建筑讲究四水归堂，上有天井，下有水景，设计师有意将室内一、二层景观相互渗透，在空间中层层相互套接，每一处好似各自独立，却又能融合成一个整体。

本案风格主调明确，与以往的茶楼概念有所不同，将传统装饰元素的经典之处，提炼并演变成新的设计符号，并运用在二楼品茶区，成为细腻且充满文化气息的细节装饰。本案通过现代简洁的设计语言来描述，拉近了这样一处充满茶香的文化空间与现代生活之间的距离。空间中的流动性、透明性、开放性以及互融性，充分体现了设计师与整个空间的艺术理念：即使身在繁杂的大都市，设计师依旧能够创造一片纯净的天空。

KEYWORDS 关键词

BRITISH STYLE
英伦风情

LIGHTING DESIGN
灯光设计

LEISURE & PLEASANT
悠闲惬意

Furnishings / Materials
软装 / 材料

TEAKWOOD
柚木

ANTIQUE BRICK
仿古砖

BLACK JADE LIGHT STONE
黑海玉透光石

LEATHER
皮革

TAWNY GLASSES
茶镜

Yincheng Red Wine Club

银城红酒会所

Designer: Li Weiqiang 设 计 师：李伟强
Design Company: Newsdays Design & Construction Co., Ltd. 设计公司：集美组设计机构
Location: Guangzhou, Guangdong, China 项目地点：中国广东省广州市
Area: 330 m² 面　积：330 m²

Yincheng Red Wine Club lies in the left side of Jialisi Hotel in Shiqiao, Panyu, Guangzhou, also close to the Helenbergh noble residence. The special geographical location determines that the club can only be positioned as the high-end product, and it mainly aims at the elites of white-collar workers and business executives. In the designer's view, the positioning of the club decoration is low-key, luxury, mature, and the corresponding classical Britain style.

The design theme of the club is the city of twilight, i.e. the British gentle night. The interior of the club is decorated with delicate and exquisite wood railings, elegant and antique street lights, and unique British telephone booth. The whole space is full of British feeling, making the guests feel as if they were in the London's streets. In addition, all kinds of world famous paintings of different sizes hung on the wall are selected carefully, quietly showing the guests their history and story. The facade of the two-story wall is divided into lattices of different sizes by the wooden lines, which are respectively set with grey lens and LED screens. People's singing and laugh, and the gorgeous clothing and fragrance are mirrored in the grey lens; foiled by the tea grey and soft light, it looks like a fresh classical painting; at the same time, it echoes with the theme paintings on the surrounding wall, statically and dynamically, falsely and truly, further strengthening the interest and cultural taste of the space; LED screens with fashion elements make the guests closer to the space, and the inverted images on the background wall bring people dream feeling.

When the night falls, exhausted people come here to enjoy the leisure of this city. With the designer's elaborate design, the flare attracts the guests to Yincheng, which seems comfortable, quiet and tasteful at this moment. In the light and slow music, the soft and clear light (high color rendering of light source works with the dimming system) creates light spots on the desk to satisfy the needs of lighting, and also build an elegant and slightly mysterious atmosphere, like a gentleman's mind, gentle and charming. When several friends sip cups of wine while enjoying masterpieces on the wall or chatting freely, they will feel leisure and easy from the light and shadow in the cups.

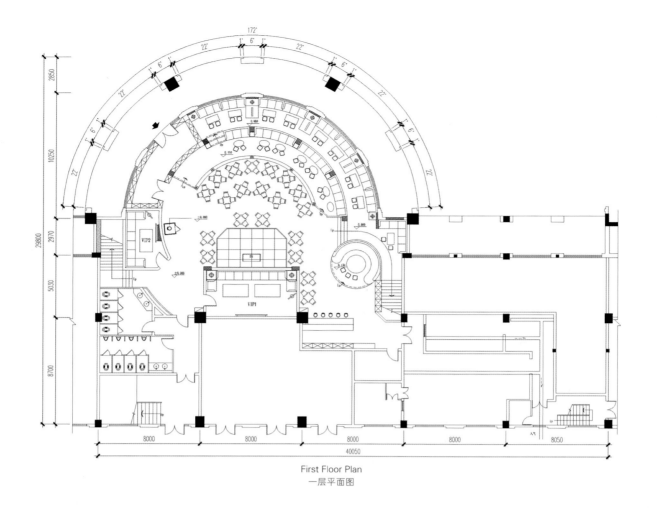

First Floor Plan
一层平面图

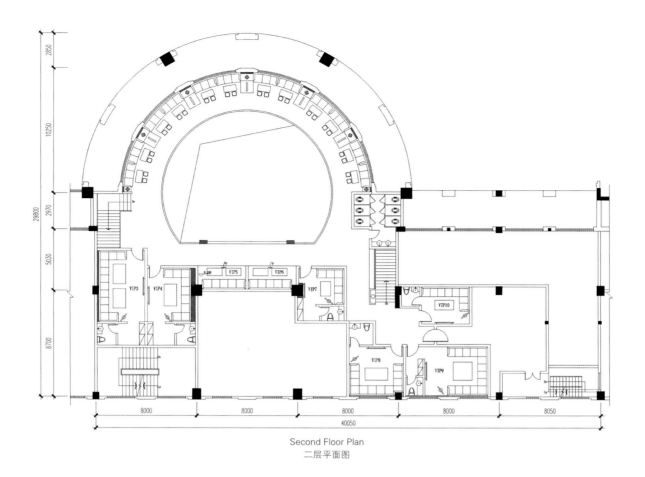

Second Floor Plan
二层平面图

银城红酒会所位于广州番禺区市桥嘉立思酒店左侧，临近海伦堡等高端住宅区。特殊的地理位置决定了此会所只能定位为中高端，其所针对的客户群是以白领企业高管为主的精英人士。设计师对会所装饰的定位是低调、奢华、成熟，而与之相对应的风格为古典英式。

该会所的设计主题为暮光之城——英伦绅士之夜。会所内部采用精致细腻的实木栏杆、优雅华贵的仿古路灯、风格独特的英式电话亭装饰，处处洋溢着浓厚的英伦风情，令客人仿佛置身于伦敦街头。此外，周边的墙上挂满了大小不一的世界名画印刷品，每幅画都经过精挑细选，安静地向客人讲述着它们各自的历史与故事。酒吧两层楼高的主墙立面用木线分割为大小不一的方格，分别镶嵌茶灰镜和 LED 屏幕，茶灰镜把大厅的怀旧景象以及酒吧内人们的欢歌笑语、衣香鬓影尽收其中，

在茶灰底色和柔和光线的过滤下，好似一幅鲜活的古典风情画；同时与周边立面上的主题绘画形成"一动一静、一虚一实"的呼应，进一步加强了空间的趣味性和文化品位；而几个LED屏幕则添加了时尚元素，拉近了客人与空间的距离，透过热闹的酒吧凝视背景墙上的倒影，给人恍然若梦般的感觉。

夜幕降临后，卸下一身疲惫的人们在这里享受着都市的悠闲。在设计师的精心设计下，一片片光斑吸引客人走进银城。此刻的银城舒适而幽雅，在轻缓的音乐中灯光柔和而清晰（高显色性的光源配合调光系统），聚集在每个桌面上的光斑既满足了照明的需求，同时也营造出优雅而又略带神秘的氛围，如同绅士的胸怀——温雅而迷人。置身其中，三五知己一边细品杯中的佳酿，一边欣赏墙上大师们智慧的结晶，或是天南地北畅所欲言，悠闲惬意尽在杯中光影之间。

KEYWORDS 关键词

JIANGNAN CHARM
江南韵味

SPACE CONCEPT
空间概念

SENSE OF LUXURY
奢华感

Furnishings / Materials
软装 / 材料

CHINA BLACK
中国黑

STAINLESS STEEL
不锈钢

GREY LENS
灰镜

LASER CRAFT GLASS
激光工艺玻璃

LIGHT STONE FILM
透光石胶片

LED LIGHT SOURCE
LED光源

BLACK MIRROR STEEL
黑镜钢

Wujiang Shengze Jiacheng International Hotel Nightclub

吴江盛泽嘉诚国际酒店夜总会

Designer: Li Weiqiang
Design Company: Newsdays Design & Construction Co., Ltd.
Location: Wujiang, Jiangsu, China
Area: 3,300 m²
Photography: Li Weiqiang, Du Jiang

设 计 师：李伟强
设计公司：集美组设计机构
项目地点：中国江苏省吴江市
面　　积：3 300 m²
摄　　影：李伟强、杜江

The project is located in Shengze Town, Suzhou. Due to the low entertainment level of the whole town and the present status of local customers' extremely strong consumption ability, party A proposed that a most exclusive local entertainment facility should be built here. The designers are stuck by several key problems because of the lack of investment. First of all, what is the concept of the most high-end in local consumers' opinion? Second, how to arrange the limited funds for the most reasonable maximum utility, and how to highlight the sense of luxury of the project?

During the design process, because of the unique architectural forms, most of the buildings are of irregular shape with low utility ratio, especially easy to form dead corners. In order to solve these problems, the designer chose non-directional round as the entrance hall of the corridor, which both strengthens the tension of the space and reduces the waste caused by the abnormal space. The public space puts emphasis on the orderly geometry layout; the room design is based on the principle of adjusting measures to local conditions, using the existing venues as much as possible on the premise of meeting all sorts of use functions. The appearance of organic rooms breaks people's traditional concept of KTV rooms, and also increases entertainment and fun of the space. In addition, two silver foil reliefs standing for the silk link in the entrance hall, large wooden sculpture with water lily theme in the landscape corridor, engraved peony pattern imitating the form of Suzhou embroidery in the room, and the use of small plait purple glass, more or less show the guests unique and dense local style and features of Jiangnan " Silk City" , thus raising its cultural taste. The integration of the facade, ceiling and ground is one of the features of this case.

Through the project practice, the designer expects to show the concept of "frugal but luxury, low-key but exalted". The fact proved that luxury is not necessarily shown by gold or silver or large amounts of money. As long as the project has a good plan layout, brightly colored decoration in the key part, opportune connection point and soft adornment, it can also achieve a luxury and graceful effect.

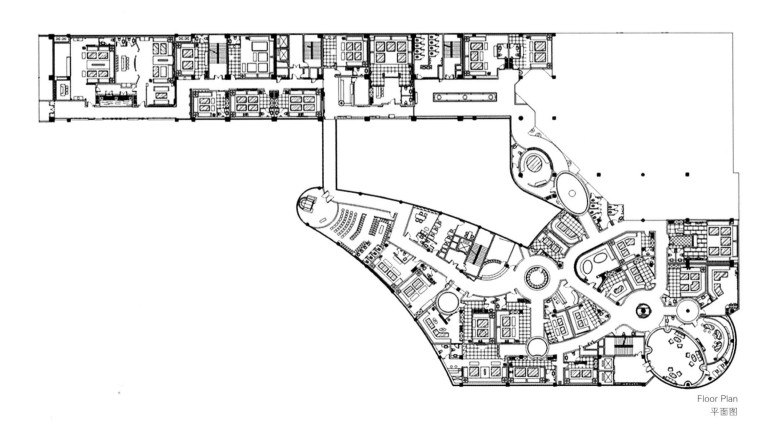

Floor Plan
平面图

项目位于苏州的盛泽镇，针对盛泽镇娱乐场所整体水平较低，而当地顾客消费能力却极强的现状，甲方提出打造一个当地最高档娱乐场所的要求。因投资有限，于是就有几个关键的问题摆在设计师面前。首先，在当地消费者心目中，最高档的概念是什么；其次，如何最合理地安排有限的资金，使之发挥最大效用，突出项目的豪华感。

在设计过程中，因为项目的建筑形态独特，大部分为不规则的形状，使用率不高，特别容易形成死角。为解决这些问题，设计师选择将若干个无方向性的圆形作为走廊的过厅，既加强了空间上的张力，也减少了因异型场地造成的空间上的浪费。相对于公共空间强调几何布局、井然有序的做法，房间设计则以因地制宜为原则，在合理满足各使用功能的前提下，尽可能利用现有场地。众多有机型房间的出现打破了传统KTV包房的概念，增加了空间的娱乐性和趣味性。此外，入口大厅中两幅表现丝绸纽带的银箔浮雕、景观走廊中睡莲主题的大型木制雕刻、房间里仿苏绣形式的牡丹图案蚀花以及小褶布纹紫色玻璃的运用，都在或明或暗、或多或少地向客人展现江南"丝绸之都"独特而浓郁的地方风貌，从而提高了文化品位。立面与天花、地面一体化也是此案的一大特色。

设计师希望通过这一项目的实践，展现"节俭而奢华，低调而尊贵"的设计理念。事实证明，表现华丽不一定要堆金砌银，只要有出色的平面布局、关键位置的浓墨重彩、连接部位点到为止，同时软装饰配合到位，同样可以达到奢华大气的效果。

KEYWORDS 关键词

RICH COLORS
色彩丰富

MODERN FASHION
现代时尚

DISTINCTIVE THEME
主题突出

Furnishings / Materials
软装 / 材料

FLOOR TILE
地砖

STONE
石材

WALLPAPER
墙纸

GLASS
玻璃

Fantasy KTV
奇幻量贩式 KTV

Designer: Zhang Xiaoying, Fan Bin, Zhu Peng, Huang Fei
Hard Decoration: DODOV
Furnishing: Root Deco International
Location: Chengdu, Sichuan, China
Area: 3,000 m²

设 计 师：张晓莹、范斌、祝鹏、黄飞
硬装设计公司：多维设计事务所
软装设计公司：诺特国际软装
项目地点：中国四川省成都市
面　　积：3 000 m²

Designed with the theme of "dream and fantasy", this KTV presents many stylish elements which are adored by today's young generation. Different rooms are designed with different themes which will cater to different customers. Designers get rid of monotonous color design and boldly use gold, silver, wine red, brown, purple and dark blue to create colorful and impressive spaces. The floor plan is another highlight of this KTV.

The materials are selected to meet the firefighting requirement, and at the same time to be energy-saving, environment-friendly and pollution-free. The floor of the rooms is paved with tiles and stone brims; the wall is decorated with wallpaper, mirrors and color paintings. The main materials need to meet the national standards and firefighting requirement, thus the wooden surfaces are treated to be fireproof. The lighting materials meet the illuminating requirement, avoid light pollution and save energy at the same time. Mirrors are largely used on the wall to expand the spaces and emphasize the sense of fashion.

本案设计以梦境、奇幻这些当代年轻人追求的时尚元素为主题，符合年轻消费者的精神和心理诉求。设计采用"一个包间一个主题"的模式，打造奇幻量贩式KTV，适合不同的年轻消费者。设计师摒弃了量贩式KTV单一的色调，将金、银、酒红、深咖、暗紫、深蓝等多种颜色巧妙融合，令空间色彩丰富，层次突出，具有视觉冲击感。同时，该KTV的平面设计也是一大亮点。

在选材方面，以达到消防要求为前提，始终以节能、环保、绿色无污染为中心。包间内地面采用地砖拼贴，局部石材收边，墙面采用墙纸、镜面玻璃、彩喷画等装饰。主材选用为达到国家标准和消防要求，所有木基层一律要求进行防火处理；灯光材料满足照度要求，但不造成光污染，同时满足降低能耗的要求。墙面大量使用镜子，扩大了空间，增加了层次感，也体现了现代时尚感。

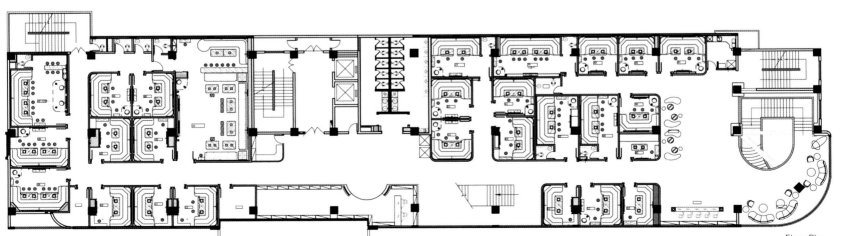

Floor Plan
平面图

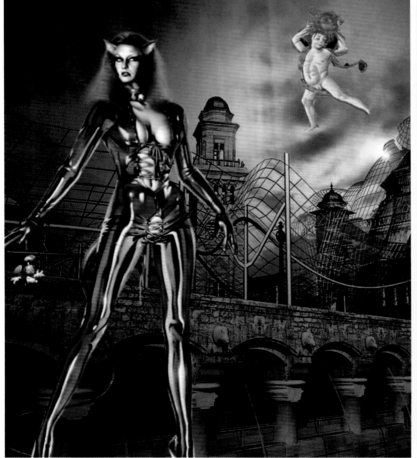

Hotel Space
酒店空间

| Cultural Integration |
| 文化融合 |

| Art Deco |
| 装饰美学 |

| Art Space |
| 艺术空间 |

| Comfortable Experience |
| 舒适体验 |

KEYWORDS 关键词

TRADITIONAL ELEMENTS
传统元素

ANCIENT STYLE
古风貌

SPATIAL AMBIANCE
空间韵味

Furnishings / Materials
软装 / 材料

ITALIAN SERPENGGIANTE
意大利木纹石

FRAXINUS MANDSHURICA TEXTURE BOARD
水曲柳肌理板

RUSTIC TILES
仿古砖

LOG
原木

LEATHER
皮革

Beijing Shouzhou Grand Hotel
北京寿州大饭店

Chief Designer: Xu Jianguo
Participating Designer: Chen Tao, Ouyang Kun, Cheng Yingya
Design Company: Hefei Xu Jianguo Architecture & Interior Design Co., Ltd.
Location: Beijing, China
Area: 16,000 m²
Photography: Wu Hui

主设计师：许建国
参与设计师：陈涛、欧阳坤、程迎亚
设计公司：合肥许建国建筑室内装饰设计有限公司
项目地点：中国北京市
面　　积：16 000 m²
摄　　影：吴辉

The project is constructed under the theme of the historic ancient city, presenting the style of ancient city south to Huai River in modern Beijing.

The elegant but simple blue bricks are applied to the space, bringing people back to the ancient age. The architecture is designed with limited height while tree roots throughout its ground floor and underground floor become the pillars, increasing the visual height, and additionally becoming the distinct mark of the traditional Anhui folk houses.

On the basis of traditional "form", the designers adopt modern material—marble for the black columned stylobate and the cream-colored pillars. The integration of modern material and traditional form leads to unique effect.

位于北京的寿州大饭店以历史悠久的古城为主题所建，淮河之南的古城风貌一路北上，经设计师巧手提炼，在现代的北京演绎出别样韵味。

素雅古朴的青砖被运用在空间的很多地方，让人仿佛回到过去那个小桥流水的年代。建筑层高较低，地下一层和一层的公共区域中，设计师设置了树根贯穿两层的柱子，提升了视觉高度，同时这种安徽传统民居形式的柱子成为本案鲜明的标志。

在取传统的"形"的同时，设计师运用了现代材质来造其"实"，黑色的圆形柱基与米色柱身皆为大理石材质，现代的材质结合传统的形式营造独特的效果。

119

KEYWORDS 关键词

RECONSTRUCTION 旧建筑改造

ORIENTALISM 东方文化

TRADITION & FASHION 传统与时尚

Furnishings / Materials 软装 / 材料

TIMBER 原木

BAMBOO 竹

RATTAN 藤

WORK OF ART 艺术品

STONE SCULPTURE 石雕

Jiangsu Tianmu Brilliant SPA Holiday Hotel
江苏天目辉煌温泉度假酒店

Chief Designer: Wang Chongming
Design Company: Hangzhou Yuwang Building Decoration Engineering Co., Ltd.
Location: Liyang, Jiangsu, China
Area: 20,000 m²

主设计师：王崇明
设计公司：杭州御王建筑装饰工程有限公司
项目地点：中国江苏省溧阳市
面　　积：20 000 m²

Jiangsu Tianmu Brilliant SPA Holiday Hotel was originally an old industrial building that carries historical memory and emotional sustenance. Through research and integrated assessment, it has been transformed into an oriental SPA holiday hotel that combines the new and old, tradition and fashion, which allows guests to take a closer look at the oriental cultural deposits and feel the charm of history.

In terms of the overall design, the owner wants to use local materials and requires designers to maximize the usable area. After a series of analysis on original structure, designers make full use of the original framework, preserve the old building and increase the area threefold to meet the need. Finally, a hollow courtyard is conceived as the first step of the design work. A large number of begonia ornamental perforated windows are arranged in the first floor space to get natural light and natural ventilation, and a hollow lighting atrium is purposely situated in the middle of the second and third floor. In order to reduce the waste of resources, air bricks applied in the interior partition wall of the extension part are sprayed with colored paint, which is soundproof, cost-effective and environmental-friendly.

In order to create a sharp contrast in the lobby, classical oriental elements are adopted in the modern decoration. Materials with special textures and patterns are refined to present the extraordinary space in a modern resort hotel. It subverts tradition while exploring the oriental cultural deposits, making each view of the elegant space full of fun.

As for the guest room, every detail is tackled with great care. Timber, bamboo and rattan interplay with each other in a concise way. These materials are used to reflect the design intention to the largest extent, creating a soft, comfortable and elegant sense. In terms of color selection, different colors are used in different areas in accordance with different layouts, spaces and functions. Light color is the main color for the whole space, besides, a little bit dark color shapes a clear pattern and enriches the space dimension. In order to highlight the uniqueness of the hotel, the owner collects a lot of antiques, art works and rare stones, and places them in the key public areas, which catch guests' eyes and provide them with extraordinary beauty...

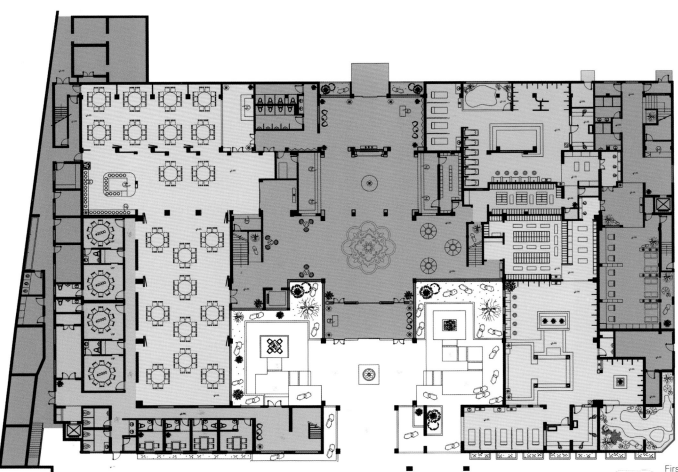

First Floor Plan
一层平面图

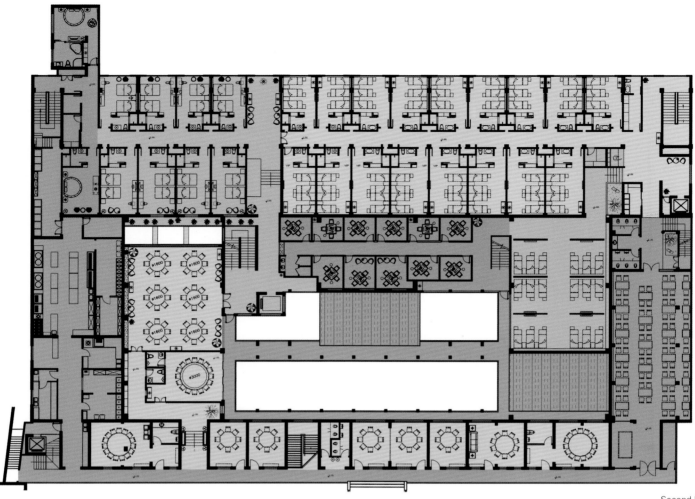

Second Floor Plan
二层平面图

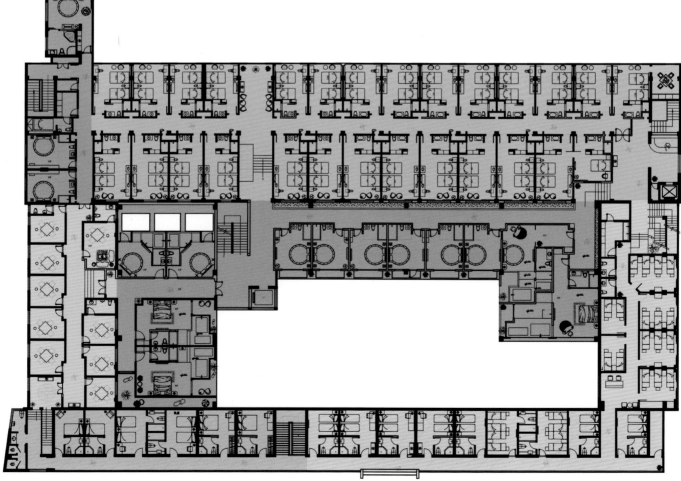

Third Floor Plan
三层平面图

江苏天目辉煌温泉度假酒店原本是一座承载着历史记忆和情感寄托的旧工业建筑，经过研究和综合评估，决定将其改造成一座新与旧、传统与时尚相结合，同时又有一定东方文化内涵的温泉度假酒店，让客人细品东方文化底蕴的同时，唤醒其对传统的记忆，感受岁月印下的痕迹，细细品读历史的变迁。

在整体的设计上，业主希望能就地取材，同时要求建筑的使用面积最大化。经过对原结构的一系列分析和研究，决定对原有的框架结构进行改造，保留旧建筑的同时还要将新加建部分的面积增加三倍才能满足项目的要求，最终设计师先以中空庭院式的框架为基础对建筑进行了改造。考虑到整体建筑内部的自然采光和自然通风，在一层设计了大量的海棠花漏窗，二、三层中间部位特意设计了中空采光天井。为了减少资源的浪费，新加建部分的室内隔墙直接采用空心砖勾缝并喷有色涂料，如此一来不但房间隔音效果好、节约成本，还节能环保。

在酒店大堂的设计上，为了达到鲜明的对比效果，在现代装饰中巧妙贯穿了东方古典元素，运用材料特有的质感和图案来演绎现代度假酒店的非凡空间。在挖掘东方文化底蕴的同时，颠覆传统，使典雅的空间中每一处景致都充满趣味性。

在酒店客房的设计上，从细节入手，整个空间装饰采用了原木、竹、藤等材料，手法以简洁为主。材料选用以最大限度体现设计意图为原则，做到少而精，统一中求变化，营造出柔和舒畅、高贵时尚的感觉。在色彩的应用上，根据格局、空间、功能的不同而有所区分，整体以浅色为主，利用重色对比使格局分明，丰富了空间层次；为了突出酒店的独特性，业主搜罗大量的古董、艺术品、稀有石雕等摆放在重要的公共区域，使之成为焦点，目的是打造不平凡的美学触觉，让人怦然心动，久久难忘……

KEYWORDS 关键词

MODERN ELEMENTS
现代元素

FOLK CUSTOMS
民族风情

POETIC & GRACEFUL
诗意优雅

Furnishings / Materials
软装 / 材料

GRID CHANDELIERS
方格吊灯

STONE
石材

GLASS
玻璃

STAINLESS STEEL
不锈钢

WOOD
木材

CRYSTAL BEAM
水晶束

Radisson Blu Hotel Liuzhou
柳州丽笙酒店

Design Company: G.I.L. Art & Design Consultants
Location: Liuzhou, Guangxi Zhuang Autonomous Region, China

设计公司：上海 G.I.L.
项目地点：中国广西壮族自治区柳州市

Located in the central business district, Radisson Blu Hotel Liuzhou overlooks the scenic Liujiang River and the municipal square across the street, and boasts superior advantages no matter in geographic location or in facilities. Fashional modern elements and unique folk custom culture play the leading roles in the hotel, besides, distinctive regional tapestry and silver background start a trip to explore the interior space.

The lobby stair, which connects Zest Bar, Globe Restaurant on the second floor and Ju Long Xuan Restaurant, becomes a corridor of time and space, and interplays with the well-proportioned chandeliers and waterscape.

Designers did something quite novel for Zest Bar, creating a warm, sweet and comfortable atmosphere. Bake House, which provides all kinds of homemade and delicious baking cakes, has the style of its own. Global Restaurant is natural and stylish, in which the open kitchen relaxes the diners, and stone, glass, stainless steel and wood are the main decorative materials here, creating a natural world. In Ju Long Xuan Restaurant, the ceiling is also skillfully designed, the translucent large-scale light box, stainless steel tendril and crystal beam largely improve the interior quality, and enrich visual and spatial effect, what's more, the Chinese butterfly patterns on the lighting in the entrance area are elegant and poetic.

The 630 m² multi-functional banquet hall provides a beautiful and exotic picture for the guests. A star ribbon on the ceiling of the swimming pool reaches to the sky, creating a bright and quiet space atmosphere. Starry ceiling and blue mosaic bottom of the pool add radiance and beauty to each other, so that the guests can't help but jump into the pool to enjoy the tender water.

柳州丽笙酒店坐落在风景迤逦的柳江河畔东岸中心商务区，与市政广场隔街相望，无论在地理位置还是设施配套上均占尽优势。整个酒店以时尚的现代元素与独具特色的民族风情文化为主导，以特有的地域织锦、银饰背景，开启了酒店探秘之旅。

连接着首层的泽斯特吧与二层的环球餐厅，聚龙轩餐厅的大堂楼梯成了时空走廊，与错落有致的方格吊灯和大堂水景交相辉映。

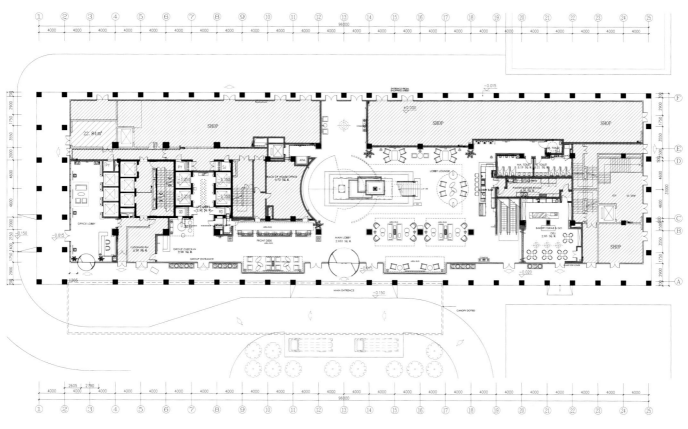

1/F Public Space Fixture & Furnishing Key Plan
一层公共部分家私总平面图

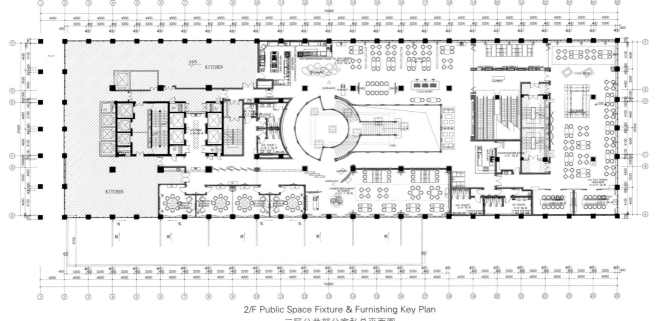

2/F Public Space Fixture & Furnishing Key Plan
二层公共部分家私总平面图

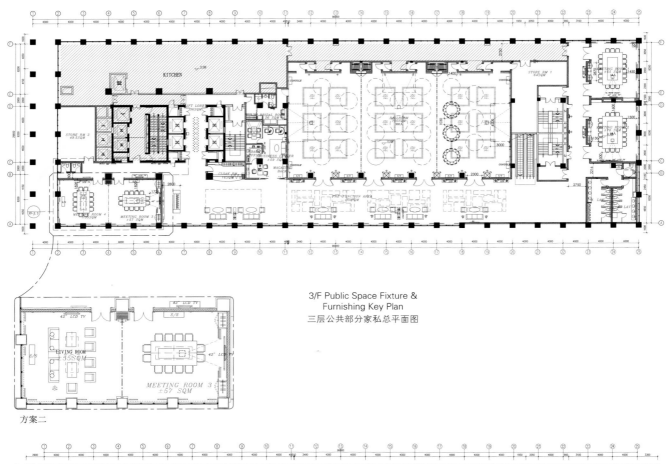

3/F Public Space Fixture & Furnishing Key Plan
三层公共部分家私总平面图

方案二

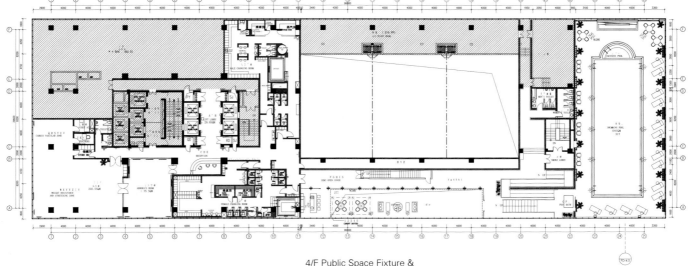

4/F Public Space Fixture & Furnishing Key Plan
四层公共部分家私总平面图

泽斯特吧设计新颖，温馨舒适。贝克西点屋自成一体，其内供应各式自制的烘焙糕点，美味至极。环球餐厅的格调时尚自然，开放式的厨房设计让人感觉轻松，餐厅以石材、玻璃、不锈钢及木材为主要装饰材料，让人有置身大自然的感觉。聚龙轩餐厅的天花板设计同样巧妙独特，以半透明的大型发光灯箱以及不锈钢蔓条、水晶束为装饰，大大提高了室内的空间感，丰富了视觉和空间效果，中式蝴蝶图案作为餐厅入口吊灯的装饰，隐隐地透出优雅的诗意。

630 m^2 的多功能宴会厅为宾客绘制了一幅精美的、极具地域风情的画面，尽显柔美、华丽。游泳池的天花板设计犹如星光飘带，连接天际，营造出一种璀璨而又静谧的空间氛围。天花板上的点点星光与蓝色马赛克池底交相辉映，让人情不自禁地想要跃身池中，享受水的温柔。

KEYWORDS 关键词

CULTURAL INTEGRATION
文化融合

LOCAL CHARACTERISTICS
当地特色

ELEGANT SPACE
优雅空间

Furnishings / Materials
软装 / 材料

NATURAL OAK
天然橡木

EBONY 黑檀木

ROSEWOOD 酸枝木

CYPRESS 桧木

CARVED WHITE
雕刻白

BLACK JADE
黑玉石

DANISH GRAY MIRROR
丹麦灰镜

Pan Pacific Ningbo Hotel
宁波泛太平洋大酒店

Chief Designer: Jiang Xiangyue
Participating Designer: Xu Yunchun, Wang Peng, Zhao Xiangyi
Area: 85,000 m²

主设计师：姜湘岳
参与设计师：徐云春、王鹏、赵相谊
面　　积：85 000 m²

Pan Pacific Ningbo Hotel is funded by Ningbo Government and operated by Pan Pacific Hotels Group. It is a typical business hotel with large area and complete functions. The designers consider not only the combination of Western and Chinese culture but also the typical style of Pan Pacific hotels as well as the unique local culture of Ningbo City. Each part of the hotel is designed with different cultural connotation. For example, there is romantic and mysterious Italian restaurant, open and spacious cafeteria, and Chinese restaurant full of traditional flavor. All these cultures and emotions are combined and presented in an elegant way.

该项目由宁波市政府出资、新加坡泛太平洋管理集团管理，属于典型的城市商务酒店，面积较大，功能较全。设计上除考虑中西文化的结合之外，还兼顾了泛太平洋酒店惯有的气质及宁波当地独特的文化底蕴等多种要素。每一个分部空间都因其特殊的性质被赋予了不同的文化精髓，如浪漫神秘的意大利餐厅、通透开敞的自助餐厅、传统典雅的中式餐厅等。东西方文化及众多情感要素在空间中的融合均通过优雅的方式进行展现。

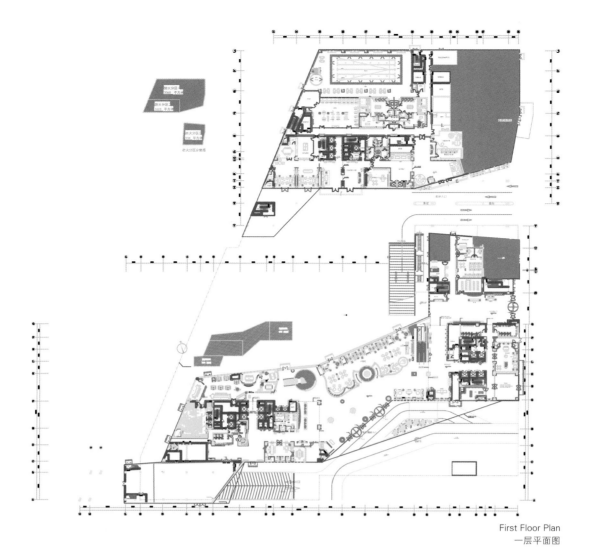

First Floor Plan
一层平面图

KEYWORDS 关键词

CHINESE STYLE
中式风格

SIMPLE & ELEGANT
简洁典雅

POETIC SPACE
诗意空间

Furnishings / Materials
软装 / 材料

FRAXINUS MANDSHURICA VENEER 水曲柳木饰面

SESAME GREY GRANITE 芝麻灰花岗岩

BLACK BRICK 青砖

RUSTIC TILE 仿古地砖

Xianghe Bainian Hotel
祥和百年酒店

Chief Designer: Xu Jianguo
Participating Designer: Chen Tao, Ouyang Kun, Cheng Yingya
Design Company: Hefei Xu Jianguo Architecture & Interior Design Co., Ltd.
Location: Hefei, Anhui, China
Area: 1,600 m²
Photography: Wu Hui

主设计师：许建国
参与设计师：陈涛、欧阳坤、程迎亚
设计公司：合肥许建国建筑室内装饰设计有限公司
地　　点：中国安徽省合肥市
面　　积：1 600 m²
摄　　影：吴辉

With concise design techniques in Chinese style, designers realized the perfect combination of restrained and unrestrained temperaments. Interspersed Chinese techniques used in this design seem unobtrusive, but present an eclectic beauty. Through transforming the classical vocabulary geometrically, visually, contrastively and rhythmically, the lines in design are simplified; stable and noble color has become the only option, and solemn, luxury and even "straight to the point" turn out to be the keywords in spreading space language.

Exterior facade that can be seen in Chinese garden is delightful at the first sight, and the screen wall in the entrance area presents a picture that a winding path leads to a secluded quiet place and the landscape is changing constantly. The atmosphere in the interior space is also quite proper; overall rhythm, materials and colors are well arranged. In addition, the element of Twelve Girls Band is lively presented.

Based on the local culture, designers combined Huizhou culture with Chinese culture and described the hotel in various ways, and some parts of the design are the clips from Huizhou culture. They cited "Orchid Pavilion" to show the artistic conception and create poetic space, and strived to break the current situation to find China's own style, so that people would truly understand the concise, simple and refined essence of Chinese design style. In some boxes, dining area and lounge area are separated reasonably, and there is a high patio in the lounge area where guests could talk freely, and enjoy the music and the moon.

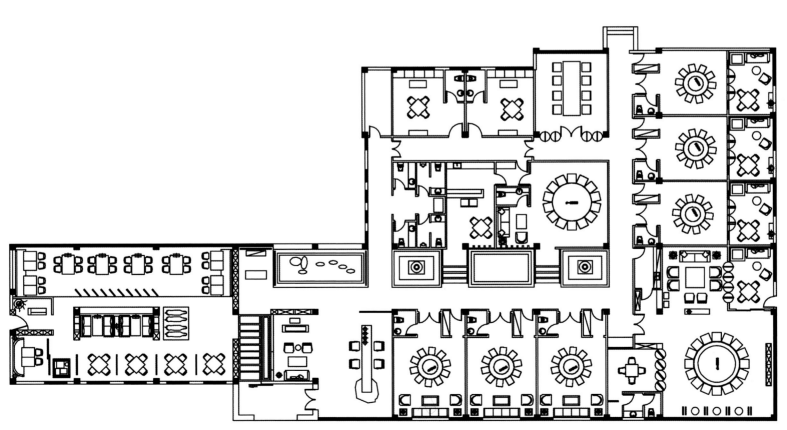

First Floor Plan
一层平面图

本案以简洁的中式手法，将含蓄内敛与随意自然两种气质完美结合。中式简约的风格在设计中得以穿插运用，不显突兀，反而呈现出一种兼容并蓄的美。通过把所谓的古典语汇几何化、图像化、对比化、节奏化，在线条上化"繁"为"简"；在色调上，采用稳重而贵气的单一色彩；在空间语言传播上，主张厚重、庄严、奢华，甚至是"开门见山"。

外立面采用中式园林的手法，给人眼前一亮的感觉。入口则用了照壁的方式，将曲径通幽、移步换景的意境表达了出来；室内也控制得相当得当，整体的节奏以及材料和色彩都把握得很到位，女子十二乐坊的元素运用得很生动。

设计师立足中国本土文化，把徽州文化与中式文化完美结合，运用了多种表现形式。设计的某些部分是徽州文化的剪辑，设计师借用《兰亭序》，表现了本案的设计意境。在本案中，设计师着力打破设计现状，找到属于中国人自己的创新的设计风格，打造具有诗人情怀的空间，让人们真正了解中式设计风格的简洁、简约、儒雅。在部分包厢设计中，就餐区与休息区合理分开，休息区设有很高的天井，这样的环境适合大家畅所欲言，听歌赏月。

KEYWORDS 关键词

EUROPEAN STYLE
欧式风格

VISUAL ARTS
视觉艺术

COLOR MOTIF
色彩基调

Furnishings / Materials
软装 / 材料

GREY MIRROR
灰镜

BLACK STAINLESS STEEL
黑色不锈钢

CASTLE GREY STONE
古堡灰石材

CHINESE BLACK STONE
中国黑石材

ART PRINTING GLASS
艺术印刷玻璃

GREY SYNTHETIC LEATHER 灰色人造皮革

Chongqing Weisilai Xiyue Hotel

重庆威斯莱喜悦酒店

Designer: Lai Xudong, Xia Yang
Design Company: Chongqing Niandai Creation Interior Design Co., Ltd.
Location: Chongqing, China
Area: 7,500 m²
Photography: Sun Huafeng

设 计 师：赖旭东、夏洋
设计公司：重庆年代营创室内设计有限公司
项目地点：中国重庆市
面　　积：7 500 m²
摄　　影：孙华峰

Located in the national pilot Liangjiang New Area in Chongqing, Weisilai Xiyue Hotel grew out of Weishilai Hotel, which was once a government office building. The current owner expects innovative design to create a brand new and competitive mid-range business hotel.

Designers aim to set off the art scene by black, white and grey, giving this modern and characteristic western romantic hotel an added unique artistic temperament, so as to let the guests enjoy a brand new hotel experience and feel satisfied here.

In terms of building appearance, the original popular cream color is replaced by calm blue and gray, creating a complementary relationship with warm yellow light at night. Designers say no to ordinary when selecting furnishings, and they use cutting-edge elements to echo with hotel appearance and present an international fashion visual art show. The typical western romantic curve modeling, western modern furniture, lighting, building sculpture and photographs echo with the European style appearance. Black, white and grey are used as the keynote to set off unique art works and furnishing. In addition, bright green, purple and orange are used on different floors to distinguish each other, offering different experiences.

重庆威斯莱喜悦酒店地处国家级新区——两江新区的重要路段。该项目名称原为威仕莱酒店，是由一栋政府办公楼改建而成。现业主希望通过设计，打造一个全新的、有竞争力的中端商务酒店。

酒店定位为具有西方浪漫主义现代风格的、个性鲜明的时尚酒店空间，以黑、白、灰来烘托其艺术品位，拥有独特的艺术气质，让顾客在享受全新的酒店体验的同时感到物超所值。

建筑外观上，将原有通俗的米黄色改为冷静内敛的蓝灰色系，与夜晚暖黄的白炽灯光形成淡淡的补色关系。陈设品的运用拒绝平庸，展现出前沿个性的姿态，在呼应酒店外观的同时也成为国际化、时尚化的视觉艺术秀场。设计通过西方典型的浪漫主义曲线造型和西方特色的现代家具、灯具以及西方建筑雕塑、摄影画与酒店的欧式外观相呼应。整体以黑、白、灰色系为基调，烘托独特的艺术品、陈设品，同时客房部分加入亮丽的绿、紫、橙色系来划分各楼层，以强调色彩对比度，并带来不同的入住体验。

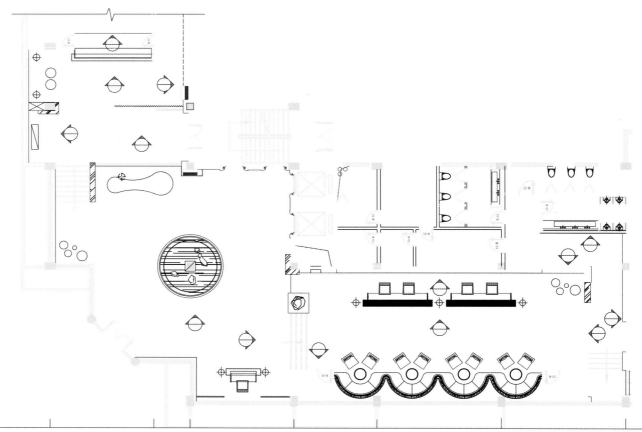

Floor Plan 平面图

KEYWORDS 关键词

ELEGANT & GENEROUS
优雅大气

RESTRAINED & SIMPLE
内敛简约

CULTURAL DEPOSITS
文化底蕴

Furnishings / Materials
软装 / 材料

BLUE FLAGSTONE
青石板

RUSTIC TILE
仿古砖

EMULSION VARNISH
乳胶漆

SQUARE TUBE
方管

DISTRESSED CEDARWOOD
杉木做旧

Juchunyuan Boutique Hotel
聚春园驿馆

Designer: Jin Shuyang, Liu Guoming, Chen Jianying, Li Hong, Wang Qifei, Cai Jiaquan, Zhang Huijing, Yu Feng
Approved by: Ye Bin
Design Company: Fujian Guo Guang Yi Ye Decoration Group
Location: Fuzhou, Fujian, China
Area: 4,000 m²

设 计 师：金舒扬、刘国铭、陈剑英、李宏、王其飞、蔡加泉、张慧晶、余峰
方案审定：叶斌
设计公司：福建国广一叶建筑装饰设计工程有限公司
项目地点：中国福建省福州市
面　　积：4 000 m²

This hotel was a courier station in ancient times, which is located in the Three Lanes and Seven Alleys, a building community that can best represent the centuries-old culture of Fuzhou. In the beginning of design work, the designers "act upon the overall pattern of this community and extract a design concept that surpasses the original one". They use natural design on the original building and enrich the Chinese style.

The courier station in ancient times was used by riders – who have urgent business (delivering government documents) – to exchange exhausted mounts for new ones or have a rest. It is now a part of Chinese civilization. The present station transformed by the designers is not so unfettered and romantic as villas or luxury hotels, however, it provides a favorable place for the litterateurs. Based on the Feng-shui theory, the planning layout reflects Chinese sense of rhythm. Well-chosen lighting, tables, chairs, wardrobe and art works in Chinese style are well arranged, which are generous, elegant and simple.

Designers reserve the architectural figure of the late Qing Dynasty and the early Republic of China, integrate the architectural features of the late 19th century, and combine them with modern elements, so as to reach a perfect incorporation of oriental charm and western elegance, creating a unique atmosphere for the entire space. In conclusion, it is a boutique hotel where one can live, sightsee and enjoy oneself at will.

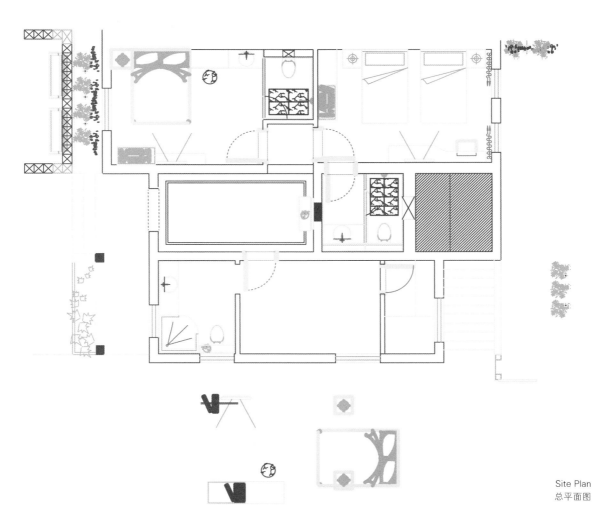

Site Plan
总平面图

本案坐落在最能代表福州悠久历史文化底蕴的建筑群落——三坊七巷之中。设计伊始，设计师就秉承"立足于坊巷整体格局特点之上，提取再整合，取之再跃之"的设计理念。设计始终坚持将设计融于自然，融于其本身独有的建筑特点之中，丰富中式建筑风格。

驿馆在古代是供传递官府文书的人途中更换马匹或休息、住宿的地方。如今驿站古迹已是中华文明的一部分。设计师理解的当代驿馆，不像别墅或豪华宾馆那么逍遥浪漫，但文人墨客在此却可借月抒怀。布局符合中国风水理论，体现了中式风格的韵律感，如淡淡的中式灯具、桌椅、衣柜以及中式艺术品，既大气优雅，又内敛简约。

该项目还有清末民初的建筑印迹，设计融合了19世纪末期建筑的特点，与一些现代元素进行碰撞，其中一些空间采用"东情西韵"的调子来诠释，中式的大气沉稳和西式的柔美优雅共处一室，使整个空间散发独特的韵味。总之，在这里可居、可观、可游、可赏，在这里可以随心、随性、随情、随景。

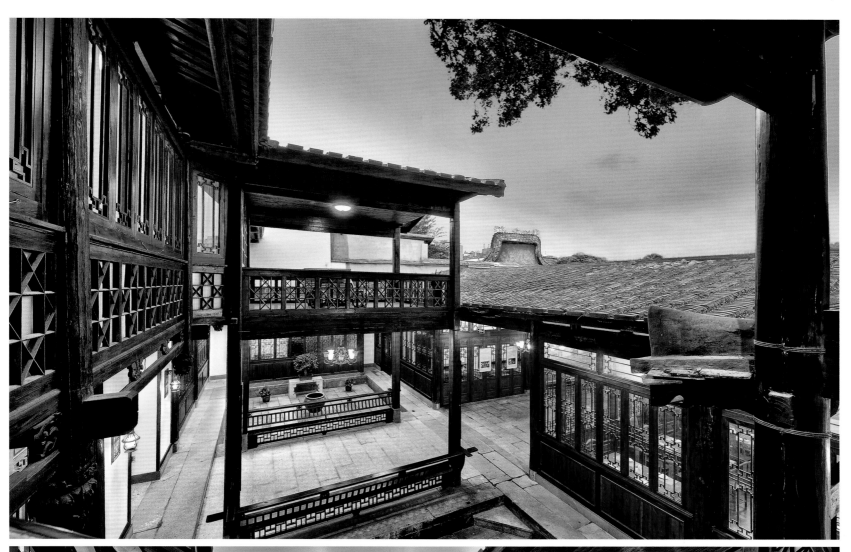

KEYWORDS 关键词
HEALTHY LIVING SPACE 养生空间
NATURAL & HARMONIOUS 自然和谐
CHINESE TRADITIONAL CULTURE 中国传统文化

Furnishings / Materials
软装 / 材料

CRYSTAL CHANDELIER
水晶吊灯

CARPET
地毯

MIRROR
镜面

WALLPAPER
墙纸

WOOD
木材

Radegast Lake View Hotel, Beijing

北京康源瑞廷酒店

Designer: Paul Liu
Location: Beijing, China

设 计 师：刘波
项目地点：中国北京市

Radegast Lake View Hotel located in Beijing is a luxurious five-star hotel that fully embraces the essence of Chinese traditional culture and the healthy living concept of Taoism. Different from the previous trend of "Zen" worldwide, "Tao" focuses more on harmony with nature and caring for life. The functional design and innovative concept of the hotel perfectly reflect the fundamental idea of "people first".

Apart from traditional functions of a five-star hotel, the unique offering of "Herb Cuisine" reflects traditional Chinese dietary therapy. Combining the healing attributes of the Chinese herbs with traditional food materials, and taking into consideration the needs of human body in different seasons, have been practiced in China for thousands of years.

From the chandelier in the lobby inspired by a cloud-looking Chinese herb and usage of plant patterns everywhere, to tree peony in the guest rooms, water lilies in the restaurant, decoration of bird cage and ancient musical instrument, the entire hotel space is carefully designed to embrace the classic Chinese culture, up-scale yet low-key, elegant yet humble.

位于北京的康源瑞庭酒店是一个融合了中国传统文化精髓和道家养生观念的五星级豪华商务酒店。与之前流行于国际的"禅"的概念不同，"道"的特点是与自然的完美融合和对生命的珍惜和关怀。"仙道贵生"，酒店在功能设置和创意概念上，都恰到好处地诠释了这一以人为本的观点。

除了传统五星级酒店的功能之外，独具特色的"药膳坊"就是传统中医疗食的体现。将中草药里面滋补养生的成分与食材进行完美的搭配烹调，配合人们四季不同的健康需求，这一种食用方式在中国已经有几千年的历史。

从酒店大堂的云朵状的灯饰和随处可见的植物图案的元素使用，到客房的牡丹花和中餐厅的荷花、鸟笼和传统乐器"古筝"的点缀，精心设计的酒店空间弥漫着优雅的中国传统文化的气息，含蓄、低调。

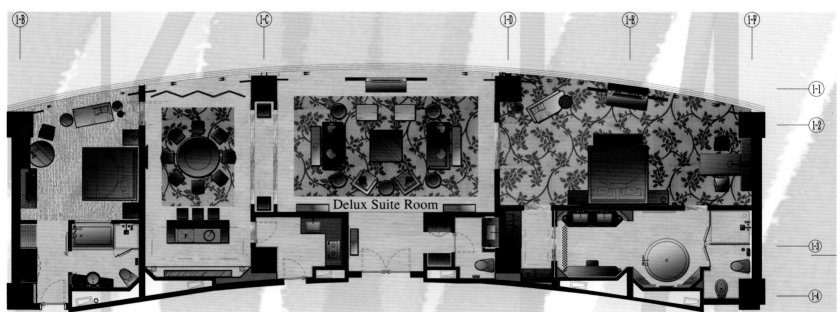

Delux Suite Room 高级套房

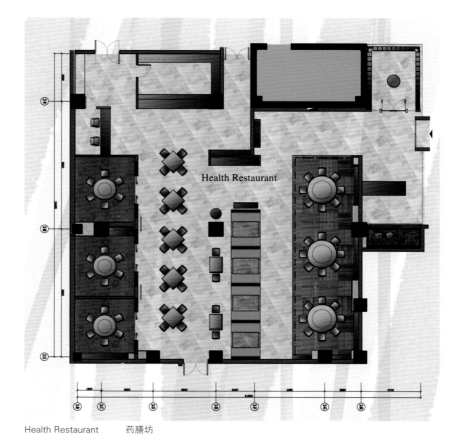

Health Restaurant 药膳坊

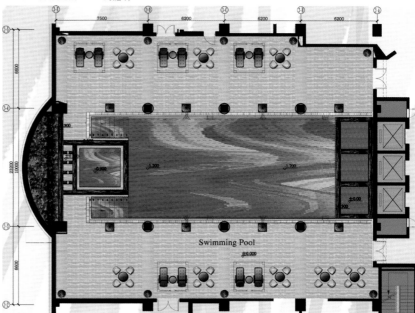

Swimming Pool 泳池

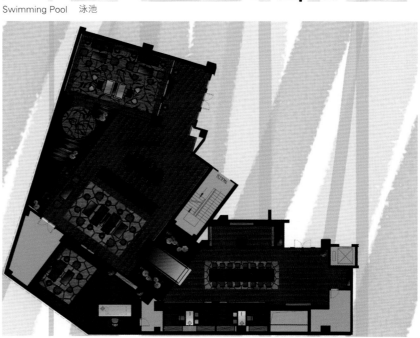

Restaurant 餐厅

KEYWORDS 关键词

SIMPLE & ELEGANT
简约典雅

HUMBLE & NATURAL
质朴自然

COMFORTABLE SPACE
闲适空间

Furnishings / Materials
软装 / 材料

WOOD
木材

RATTAN
藤艺

COTTON, LINEN
棉麻

FLANNELETTE
肌理绒布

Jintai Longyue Seaview Golf Resort, Liaoning

辽宁金泰珑悦海景高尔夫度假酒店

Designer: Paul Liu
Location: Yingkou, Liaoning, China

设 计 师：刘波
项目地点：中国辽宁省营口市

The hotel is located in Bayuquan District, Yingkou which is known as the "Pearl of the Bohai Sea". It's a park-style resort hotel and has many facilities, including guest rooms, restaurants, entertainment space, golf course, marina, SPA area, racecourse, holiday villas, beach and water sports area. For the design concept of "sun, green, sports, and entertainment," the hotel style is generally simple and elegant. The lobby, guest rooms, restaurants are all taking full advantage of the sun, sea and other natural elements. You can feel it like a heavenly place.

The building was built on the seashore with a shape like the wings of eagle, as well as ships waiting to sail, featuring a romantic atmosphere. The interior design keeps the overall style and mood in harmony with the architectural appearance: simple and elegant, humble and natural. The majestic lobby and unique shape of the walls have created a romantic atmosphere which makes people feel like in a cabin. Looking out the window, the blue sky and sea will make you feel roaming in the vast sea.

In accordance with the interior and architectural styles, brown and dark red are used to define the basic tone of the space. Some places use light beige to get bright and soft. Materials with natural texture are applied to highlight the elegance and leisure of the space. In addition, with the rattan furnishings, cotton and linen products, as well as the flannel fabrics, it enables the guests to have a fantastic holiday with birds' songs, tree shadows and surf.

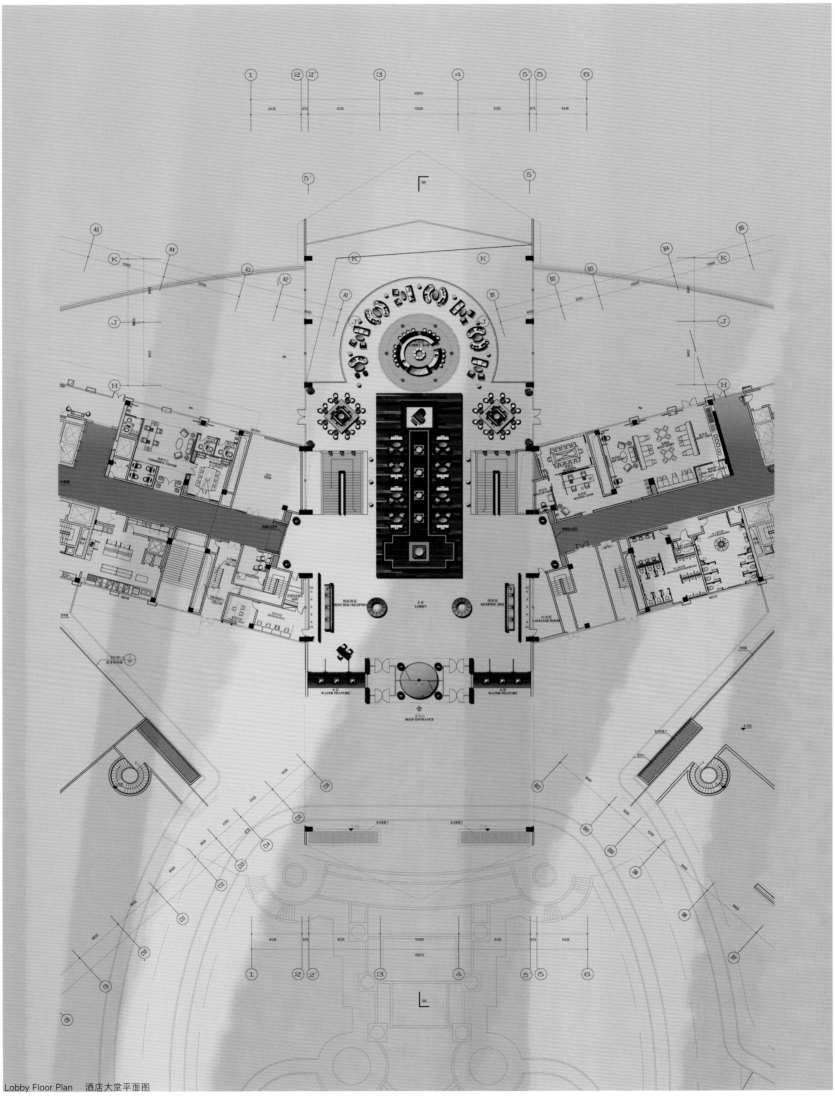

Lobby Floor Plan 酒店大堂平面图

　　金泰珑悦海景高尔夫度假酒店位于素有"渤海明珠"之称的营口市鲅鱼圈区，按白金五星级标准建造，是一个集客房、餐饮、娱乐、高尔夫、游艇码头、温泉、跑马场、度假别墅、海滨浴场及水上运动等于一体的公园式休闲度假酒店。秉承"阳光、绿色、运动、休闲"的设计理念，酒店整体风格简约典雅。酒店临海而建，布局巧妙，无论大堂、客房、餐厅均充分利用阳光、大海等自然元素，处处皆景，让宾客尽享如身临世外桃源般的惬意。

　　建筑依海而建，形态有如大鹏展翅，亦似战舰等待起航，充满浪漫而又阳刚的气息。室内设计的整体风格与建筑外观保持意境上的和谐统一：稳重大气，典雅质朴，细节处充满自然唯美的情调，与大自然的光影、冷暖完美融合。大堂气势恢宏，别出心裁的墙面造型营造出置身船舱的浪漫氛围，放眼窗外，海天一色，让人觉得仿佛遨游在一望无际的大海之中。

　　根据室内外整体格调，选用了稳重温暖的褐色、棕红色，局部点缀明快柔和的浅米色，搭配以各种天然质朴的材质。用具有自然肌理和纹路图案的物料，烘托大气、典雅、轻松且愉悦的休闲氛围，同时配以藤艺、棉麻及肌理绒布，令每位莅临于此的宾客身心放松，在鸟鸣、树影、涛声中度过美妙舒适的假日黄金时光。

Exhibition Space
展览空间

Creative Exhibition
创意展示

Spatial Esthetics
空间美学

Visual Impact
视觉冲击

Interactive Experience
互动体验

KEYWORDS 关键词

SUMPTUOUSNESS & DIGNITY 奢华尊贵

STREAMLINE 流线

EXQUISITE DETAILS 精致细节

Furnishings / Materials
软装 / 材料

DULUX EMULSION PAINT
多乐士乳胶漆

POLISHED TILE
抛光砖

ART GLASS
艺术玻璃

Huizhou Harmony World Watch Exhibition Hall

惠州亨吉利世界名表展厅

Designer: Wang Wuping
Client: Harmony World Watch Center Co., Ltd.
Location: Huizhou, Guangdong, China
Area: 220 m²

设 计 师：王五平
客　　户：亨吉利世界名表中心有限公司
项目地点：中国广东省惠州市
面　　积：220 m²

"Sumptuousness & Dignity, Remarkable Enjoyment", "less is more, less but fine" are the core theme of the project design, interpreting the status symbol. Refined sense of streamline without extravagant decorative elements, extreme colors of black and white without gold plated theme, all of the details manifest the uniqueness. The ceiling and floor under similar design, along with the cashier form the corresponding streamlines and express the inner unity of product and quality.

The exhibition hall shows exquisite design details ranging from showcases to the lighting atmosphere. The company logo on the gray glass under Maintenance Division delivers the culture and service idea of the company.

　　"奢侈华贵，非同凡享"，"少即是多，少而精"的概念既诠释了身份的象征，也是本项目设计的核心主题。没有刻意华贵的装饰元素，却有着精练的流线感；没有饰金的主题，却有着极致的黑白，一切无不彰显出其产品的独一无二。天面与地面异曲同工，与中间收银岛流线相呼应，更加传达了产品和质量的内在统一性。

　　这一世界顶级品牌手表展厅，处处体现着精致的设计细节，从展柜与背柜的设计制作，到店内整个灯光氛围的营造，无不细致入微。手表维修处下面的灰玻上，丝印着的公司的标识，也时刻传达着公司文化与服务理念。

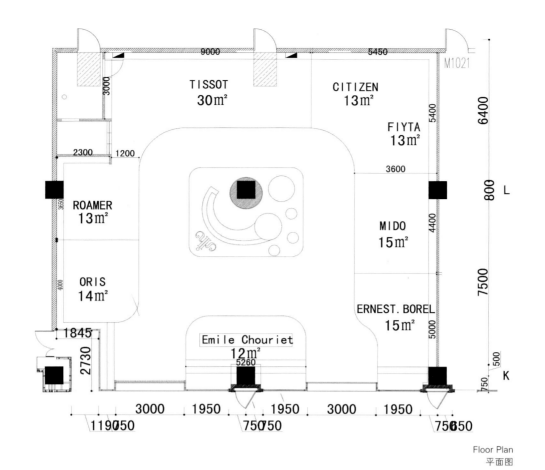

Floor Plan
平面图

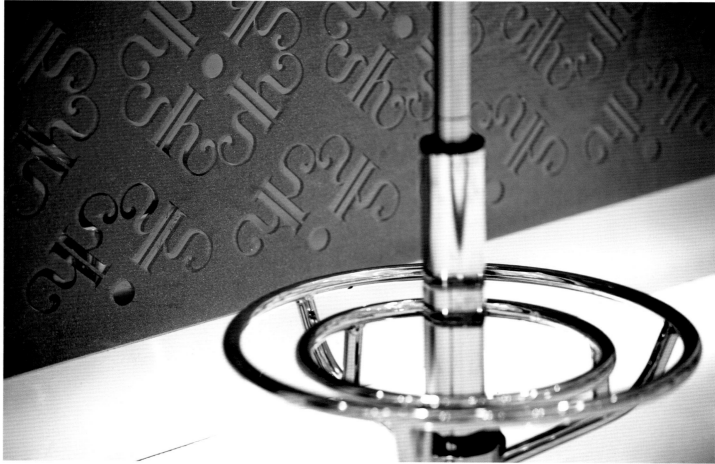

KEYWORDS 关键词

ORIENTAL ESTHETICS
东方美学

CONCISE DESIGN
简约设计

ELEGANT SPACE
典雅空间

Furnishings / Materials
软装 / 材料

EGG CARTON
蛋托

WOOD JOIST
木龙骨

CARPET
地毯

Kunshan Xupin Exhibition Hall
昆山叙品展厅

Designer: Jiang Guoxing
Design Company: Kunshan Xupin Design & Decoration Engineering Co., Ltd.
Location: Kunshan, Jiangsu, China
Area: 100 m²

设 计 师：蒋国兴
设计公司：昆山叙品设计装饰工程有限公司
项目地点：中国江苏省昆山市
面　　积：100 m²

In the limited space, a lot of fun is created by the designer. Egg cartons are adopted as the sole material in the design, manifesting Xupin's corporate spirit in a most direct way that design is everywhere and any materials in life can be applied as design elements.

Concise but not simple is the highlight of the design, and the decorative elements of clear intention, easy expression and unceasing consolidation lead the visitors to focus on the designer's intention. The films of enterprise development, furniture and accessories are set as embellishment in the space background of egg carton arrays, showing the modern Oriental esthetics. The space is mysterious, conservative, elegant and natural.

昆山会展中心的叙品展厅面积并不算大，但设计师赋予了这个小空间很多乐趣。整个设计空间只采用一种材料——蛋托，最直接地体现了叙品的企业精神——"无处不设计"，即生活中的任何材料都可以拿来作为设计元素。

简洁却不简单是这个展厅设计的亮点，意图明确、易于表达、不断强化的装饰元素时刻引导着观众专注于设计师的意图。在序列蛋托的空间背景下，点缀着企业发展业绩胶片、家具以及饰品，从容展现现代东方美学，赋予空间扑朔迷离、涵养深湛、儒雅高贵的气度。

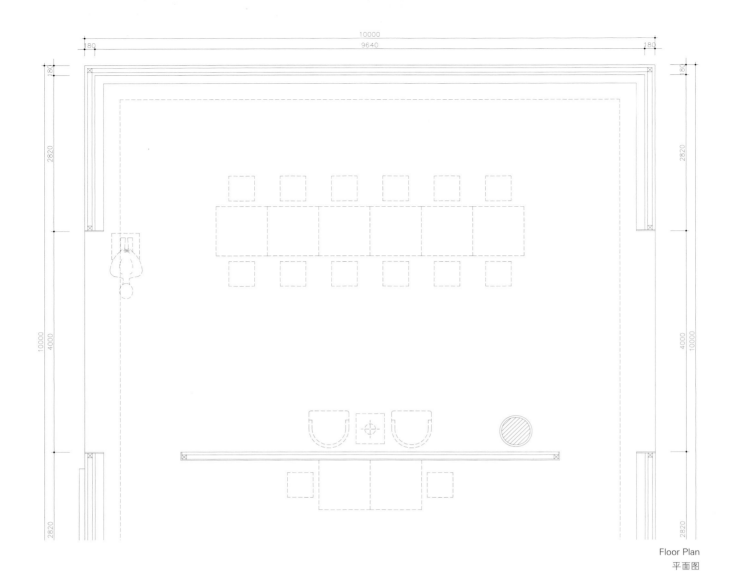

Floor Plan
平面图

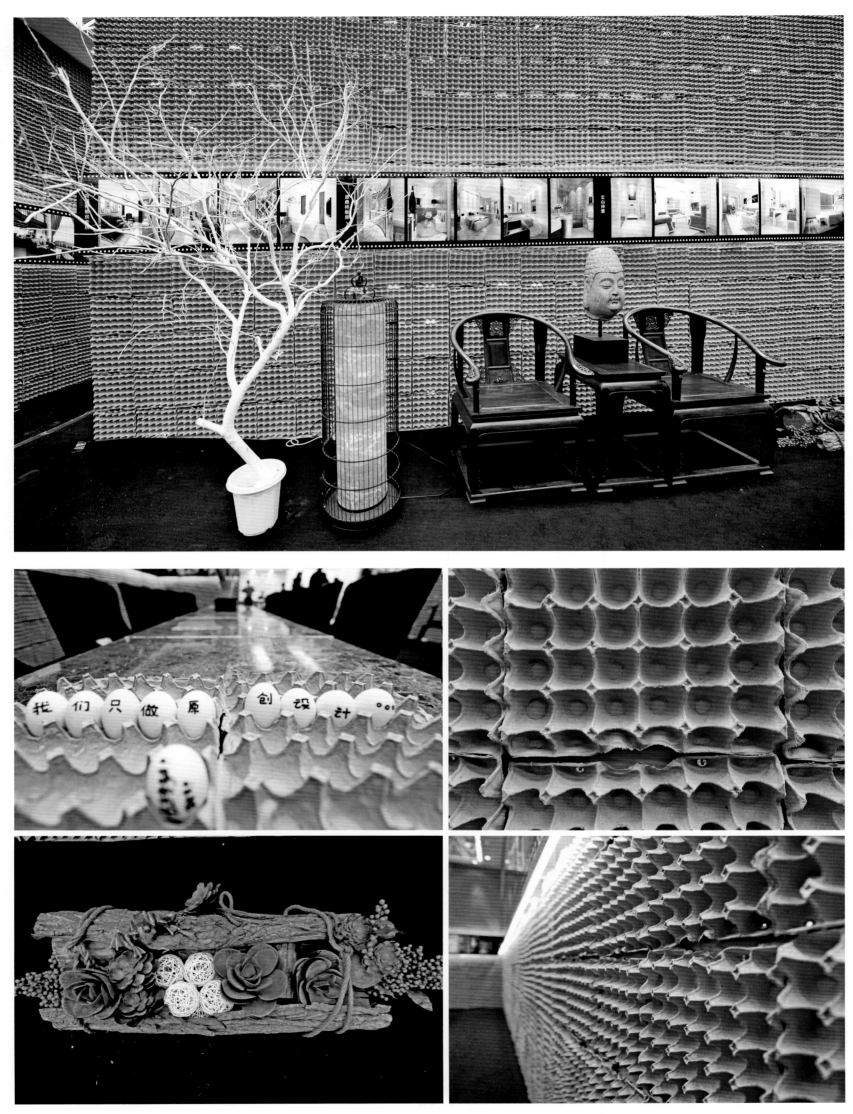

KEYWORDS 关键词

SENSE OF WHOLENESS
整体感

DECORATIVE COLLOCATION
装饰搭配

EXHIBITION SPACE
展示空间

Furnishings / Materials
软装 / 材料

HYDROGEN FLUOROCARBON PAINT FOR SQUARE TUBES
方通氟碳氢漆

WHITE TEXTURAL COATING
白色质感涂料

TEMPERED GLASS
钢化玻璃

SELF-LEVELING CEMENT FLOOR 水泥自流平地面

AGED STEEL PLATE
做旧钢板

HORSE HIDE
马皮

Phase II Exhibition Hall of Shenzhen O'seka Art Exhibition Center

深圳奥斯卡艺展中心二期展厅

Designer: Yao Haibin
Design Company: Yanshe Design
Location: Shenzhen, Guangdong, China
Area: 1,000 m²

设 计 师：姚海滨
设计公司：深圳市砚社室内装饰设计有限公司
项目地点：中国广东省深圳市
面　　积：1 000 m²

As a carpet exclusive store, the project is located on the seventh floor of Art Exhibition Center (Phase II) in Luohu District, Shenzhen. With great attention to the general decoration of the façade, the designer arranges technically settings and props with background decoration, proper lighting, colors and patterns to conduct the product introduction and promotion. More than product exhibition, the design of interior exhibition area also pays attention to the smooth traffic circulation.

深圳奥斯卡为地毯专卖店，位于深圳罗湖区的艺展中心二期7层。设计师在橱窗的布置上特别注重门面总体装饰，巧用布景、道具，以背景装饰为衬托，配以合适的灯光、色彩和实物图案，进行商品介绍和商品宣传。室内展示区在充分展示商品的同时，保证客流的交通流畅。

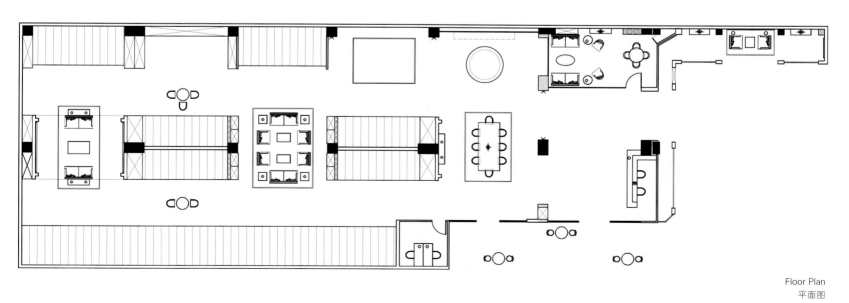

Floor Plan
平面图

KEYWORDS 关键词

COLOR
色彩

CONCISE DECORATION
简约装饰

DISPLAY SPACE
展示空间

Furnishings / Materials
软装 / 材料

RECYCLED AGED WOOD
回收老木板

OAK
橡木

PINE CARBONIZED PLATE
松木炭化板

ITALIAN HANDMADE BRICK
意大利手工砖

ARABESCATO CORCHIA MARBLE
雪花白大理石

Hangzhou A·Base Furnishing & Salon
杭州佰色 A·Base 陈设 & 沙龙

Designer: Zhu Xiaoming
Participating Designer: Gao Liyong, Lei Huawen, Zhu Lulu
Design Company: INEAR Design & Decoration
Location: Hangzhou, Zhejiang, China
Area: 1,800 m²
Photography: Lin Feng

主设计师：朱晓鸣
参与设计师：高力勇、雷华文、朱露露
设计公司：杭州意内雅建筑装饰设计有限公司
项目地点：中国浙江省杭州市
面　　积：1 800 m²
摄　　影：林峰

The project, as a multifunctional space for furniture exhibition and designer communication, combined with the characteristics of the visiting group, it presents a simple, easy and integrative space instead of miscellaneous decorative style like conventional club or show flat. A great area of pure white color is adopted in the exhibition space as the background color to highlight the products' independent form and diversity. The space is divided as BOX array to classify the products. The reception space on the second floor is accessible through the tranquil detached Designer Channel leading the visitors to change their mood. Products with dynamic colors are positioned casually in the salon area on the second floor which covers different material groups of rich texture and takes gray beige as the major color, delivering the pleasing, warm and perceptual ambiance.

该项目为一个集家具陈设产品展示、设计师交流聚会于为一体的多功能场所，结合来访群体特质，舍弃常规的"会所"、"样板房"等繁杂和较为仪式感的装饰堆砌，换以一种轻松平和、极简但富含包容性的空间气息来表现。在对外公开的家具与陈设饰品展示空间，为突显产品既有独立形态又有系列组合的多样性，色彩上采用大面积纯白色作为环境色。空间采用阵列"BOX"分区展示，合理地将产品进行了划分归类。二层的接待空间从独立静谧的"设计师通道"进入，借以转换来访者的情绪。沙龙区域则将富含肌理的不同材质进行组合，在米灰的主色中配以恰当的具有跳跃色彩的陈设产品，表达愉悦、温暖、感性的沙龙区域氛围。

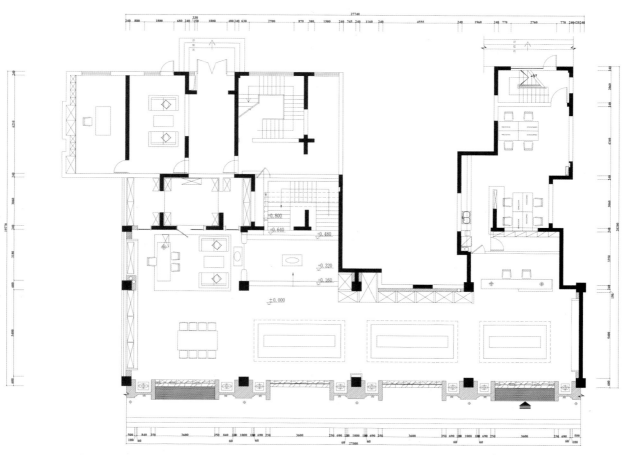

First Floor Plan
一层平面图

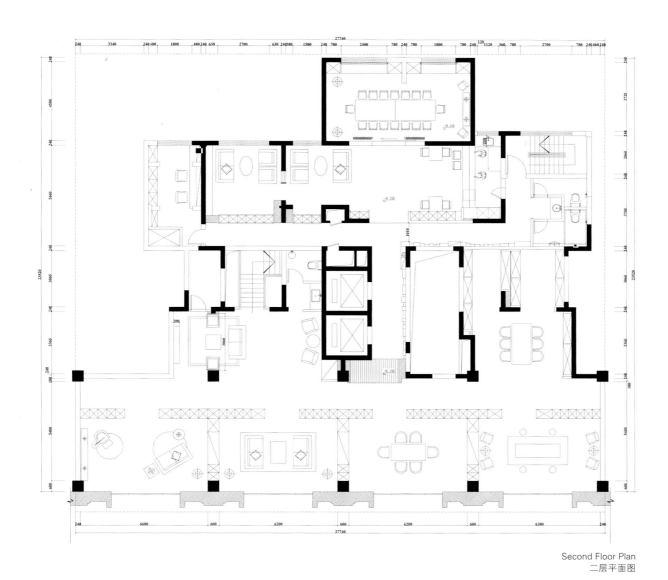

Second Floor Plan
二层平面图

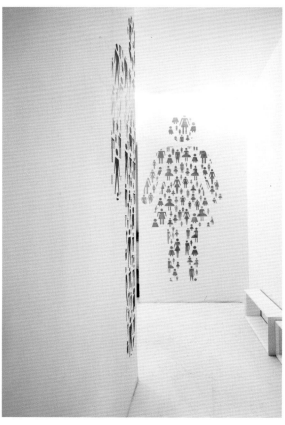

KEYWORDS 关键词

NORTHERN EUROPEAN STYLE
北欧风格

BRAND IMAGE
品牌形象

DYNAMIC DESIGN
动感设计

Furnishings / Materials
软装 / 材料

NOBELTILE
诺贝尔瓷砖

BEIGE SCAGLIOLA BRICK
米黄仿云石砖

COMPOSITE FLOOR
复合地板

CARPET
地毯

WHITE GLASS
白玻

STAINLESS STEEL
不锈钢

Wuhan Zhongda Jiangbao 4S Store Exhibition Hall

武汉中达江宝 4S 店展厅

Designer: Xia Jinsong
Design Company: Wuhan Fanshi Art Design Co., Ltd
Location: Wuhan, Hubei, China

设 计 师：夏劲松
设计公司：武汉梵石艺术设计有限公司
项目地点：中国湖北省武汉市

The project, Jiangxia BMW 4S store in Wuhan, occupies a land area of approx. 15,000 m^2, a floor area of approx. 8,100 m^2; its two floors are divided into car showroom, after-sales service hall, office zone, maintenance workshop, staff living zone, training area, etc. The design in the modern Northern European style is simple and graceful, taking white as the major color, gray and black as the comparison colors. Glass and stainless steel are applied frequently to express the advantage of high technology. The car booths are designed dynamically with reference to analog runway. Considering the brand image, the main space is designed under the corporation culture and normative application of BMW. The arrangement of lighting, plants, furniture and color collocation complies with the standard of BMW image store, delivering the graceful image and canonical sales philosophy of the brand.

该项目为武汉江夏宝马4S店，占地面积约15 000 m^2，建筑面积约8 100 m^2，分上下二层，由汽车展厅、售后服务大厅、办公区、维修车间、员工生活及培训区等组成。整体设计简洁大方，以北欧现代风格为主，色彩上以白色为主，对比有灰色和黑色，多用玻璃和不锈钢，突显高科技的优势。汽车展位模拟跑道对位展车，极具动感设计。作为品牌形象店，主体空间设计遵守宝马企业文化与规范应用，其中的光照形式、植物种植方式、家具样式、色彩搭配等都遵守宝马形象店的标准，也体现出品牌的大气形象和规范销售理念。

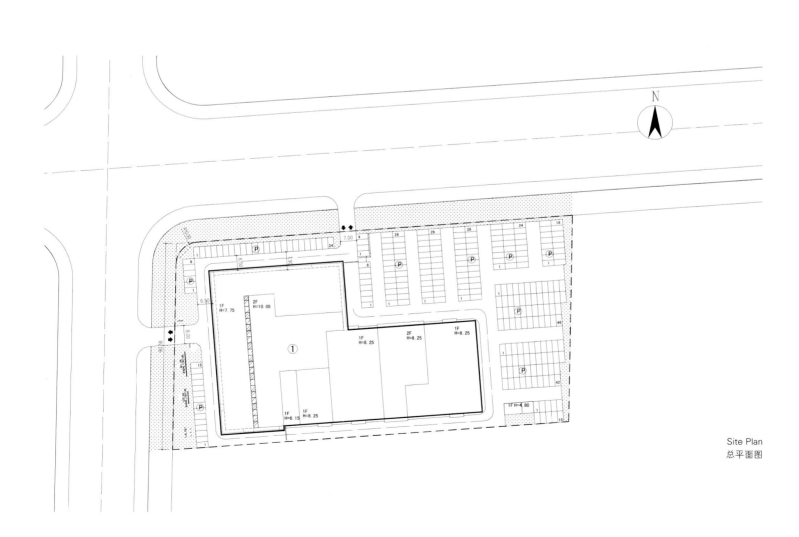

Site Plan
总平面图

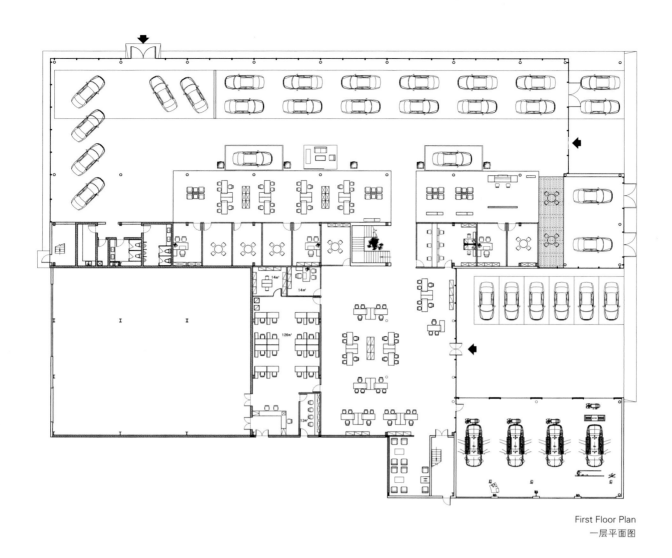

First Floor Plan
一层平面图

Second Floor Plan
二层平面图

KEYWORDS 关键词

MODERN & FASHIONABLE
现代时尚

COMPARISON
对比

DISPLAY SPACE
展示空间

Furnishings / Materials
软装 / 材料

TUNGSTEN STEEL
乌钢

SILVER MIRROR
银镜

GREY MIRROR
灰镜

GLASS
玻璃

LEATHER
皮革

Restonic Furniture Exhibition Stand
运时通家具展位

Design Company: Shenzhen Xima Layout Design
Location: Dongguan, Guangdong, China

设计公司：深圳西玛企划设计
项目地点：中国广东省东莞市

Designers utilize the sharp comparison between the hard and soft, the cold and warm, the dynamic and static to interpret a modern, fashionable, romantic and fantastic exhibition space.

The square-shape violet mirror, leather and black steel figure like a mysterious magic cube positioned in the outside of the stand arouses visitors' infinite imagination and desire to explore the mystery. The fish-shape pendants and soft membrane beam model at the entrance look like a coquettish lady covered by light rings; the green plants on both sides embody the environmental protection trait of the products. The materials of ceiling, wall and floor create a sharp contrast in the interior space and the transparent bead curtain of the oval ceiling extends to the distance with the sound caused by gentle breeze, which sounds like the wave crash heard by people lying in bed, which is the dream and pursuit of every romanticist.

在此设计方案中设计师巧妙利用硬与软、冷与暖、静与动的强烈对比，诠释了一个不乏现代、时尚、浪漫、且梦幻般的展示空间。

在展位的外侧，方形的紫镜与皮革及乌钢造型犹如神秘的魔方，激发了观众无限的遐想与探索神秘的欲望。走近展位，入口处映入眼帘的鱼形吊坠与软膜光柱的造型犹如一个被光环笼罩的婀娜多姿的少女，而两旁的绿色植物更体现了本产品的环保特性。走进展位内部，天花板、墙面及地面材料产生强烈对比，而椭圆形吊顶的透明装饰珠帘犹如海浪般延伸消失在远处，珠帘被微风吹动撞击而发出的声音感觉像是躺在舒适的软床上聆听到的海浪拍打海岸的声音，这种美妙的感觉是每一个浪漫主义者的梦想与追求。

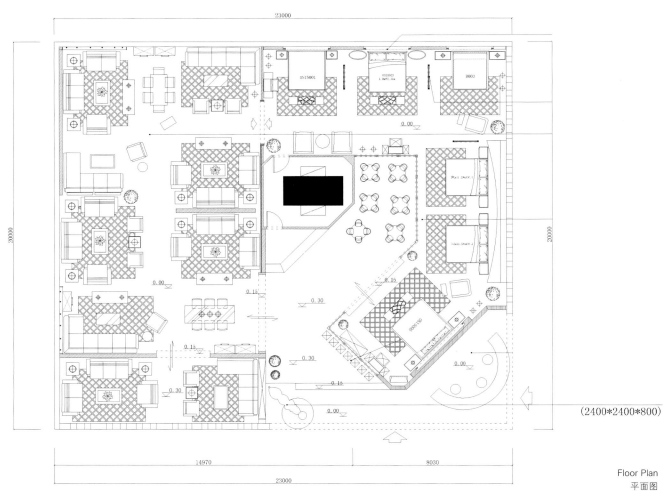

Floor Plan
平面图

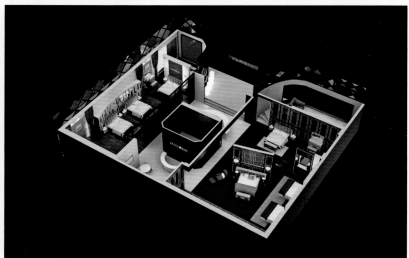

KEYWORDS 关键词

CURVED LINES
弧形线条

SOLID & VOID
虚实结合

DYNAMIC SPACE
动感空间

Furnishings / Materials
软装 / 材料

PAINT
油漆

STONE
石材

WOOD VENEER
木饰面

TILE
瓷砖

TEMPERED GLASS
钢化玻璃

Three Forks Lake Digital Display Space
三岔湖数字展示空间

Design Company: Sichuan Kevin Architectural Engineering Co., Ltd.
Client: Chengdu Tianyi Tianying Digital Media Co., Ltd.
Location: Jianyang, Sichuan, China
Area: 630 m²
Photography: X. Kevin Li

设计公司：四川凯文建筑工程有限责任公司
客　　户：成都天意天映数字科技传媒有限公司
项目地点：中国四川省简阳市
面　　积：630 m²
摄　　影：李响

Located in the Three Forks Lake Digital Center in Xinmin County, Three Forks Lake Town, Jianyang City of Sichuan Province, the space is designed with the idea of "flowing". A great number of curved lines are used to create a dynamic space to achieve a perfect combination of solid and void. It takes advantage of the regional development and the eco resources to create a display space with water and island. Modern and traditional elements are combined together to emphasize its identity and uniqueness. The design has won the 7th Shanghai "Golden Bund Award" and the second prize of "The 7th Design Expo & The 2nd International Environmental Innovation Design Competition" in 2012.

　　本案位于四川省简阳市三岔湖镇新民乡三岔湖数字招商中心内，在设计之初就确立以流动的理念，以大量的弧形线条营造出动感的空间，形成完美的虚实结合。将区域开发理念和地域生态资源巧妙结合，融入整个项目设计和氛围营造中，以水、岛屿作为空间形态的载体，让现代与传统有机结合，突出项目的唯一性和不可复制性，新颖合理。该项目荣获2012年第七届上海"金外滩"奖以及2012年第七届社博会暨第二届国际环艺创新设计大赛二等奖。

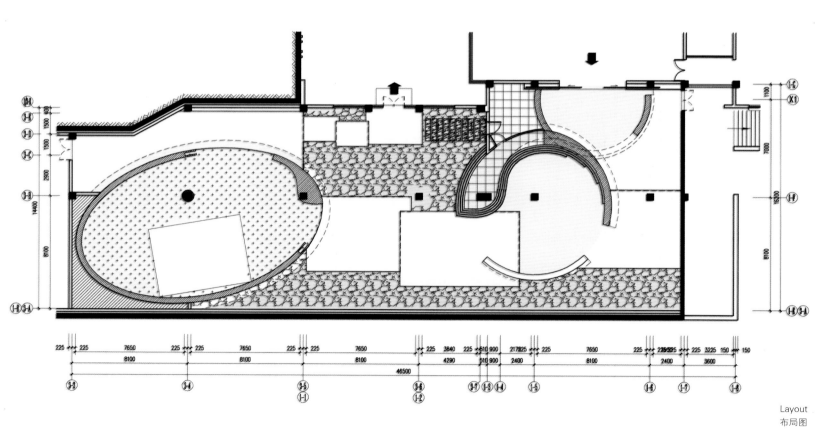

Layout
布局图

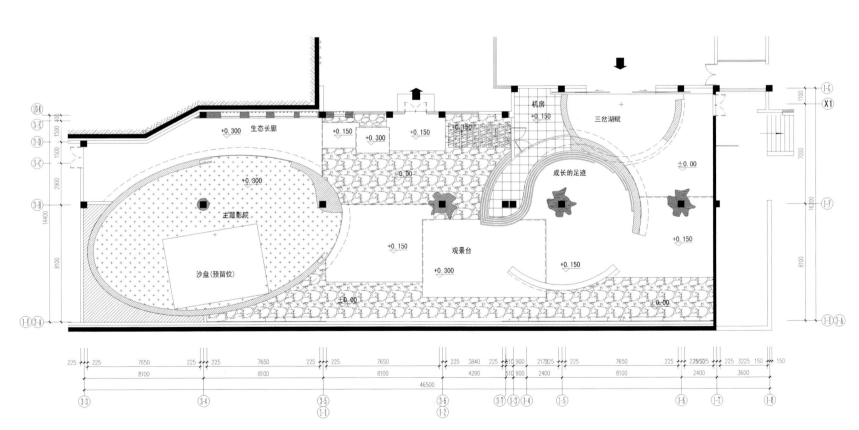

Floor Plan
平面图

275

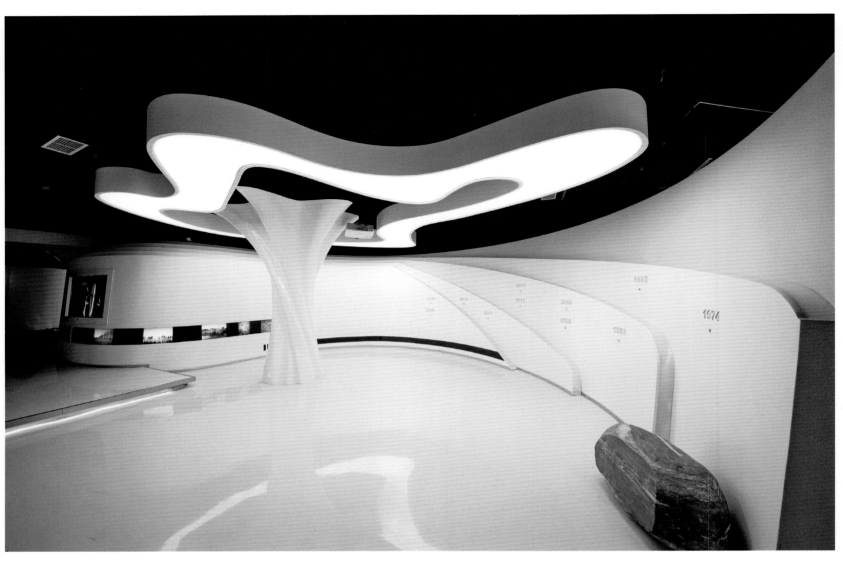

KEYWORDS 关键词

SPATIAL CHARACTER
空间特性

SENSE OF PRESENCE
存在感

MODERN MATERIALS
现代材料

Furnishings / Materials
软装 / 材料

BARE CONCRETE
清水混凝土

TAIWAN GUANYIN STONE
台湾观音石

INDIA BLACK STONE
印度黑石

SEN VENEER
栓木木皮

Original Fengjing Public Facilities

原风景大楼公共设施

Chief Designer: Su Chingchi
Participating Designer: Xie Xinwei, Zhao Huanzhen, Zhang Qunfang, Chen Yahan
Design Company: Ching-Chi Design / The Research Institute of Architecture and Interior Design
Location: Taichung, Taiwan, China
Area: 1,448 m²(Interior), 2,347 m²(Landscape)
Photography: Liu Zhongying

主设计师：苏静麒
参与设计师：谢欣薇、赵焕珍、张群芳、陈雅涵
设计公司：清奇设计 / 苏静麒建筑室内设计研究所
项目地点：中国台湾台中市
面　　积：室内 1 448 m²，景观 2 347 m²
摄　　影：刘中颖

The design under the theme of exhibition space adopts bare concrete in the main structure, and the special horizontal shading way shows the transparent sense of the space. The architecture is formally planned and set.

The wall and frame are stressed by the bare concrete, along with other modern materials like the remote and high valley. The openings in the bare concrete wall look like the grottos set into the rock wall. In the end of the corridor, a great water curtain flows from the wall and a kiosk stands before the water curtain.

　　该设计以一个展示空间的议题展开，主结构采用清水混凝土，以特殊水平遮阳的方式呈现空间的透明感，从那时起，本案的建筑规划正式诞生。

　　以清水混凝土构筑突出大地的墙与框架，像"古迹"般与其他的现代材料搭配，如同在"遥远他方的高耸峡谷"中前行。清水混凝土墙上一个个的开口像嵌入石壁上的一个个"洞窟"。廊道的尽头，只见一大片水幕自墙流入水中，有亭飘于水墙之前。

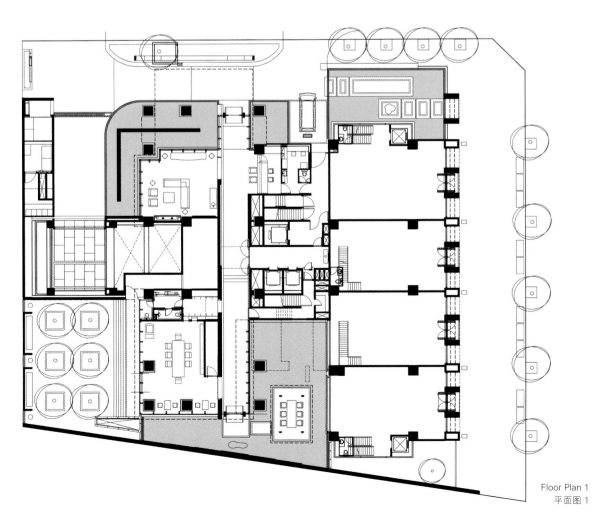

Floor Plan 1
平面图 1

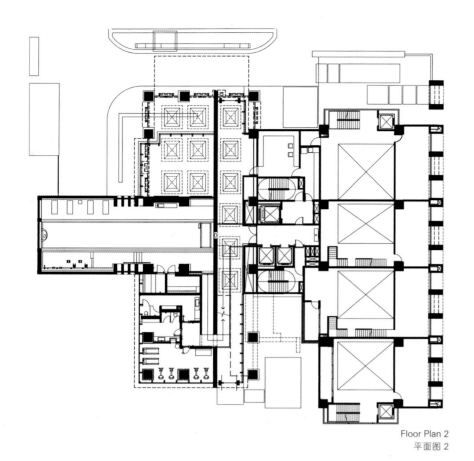

Floor Plan 2
平面图 2

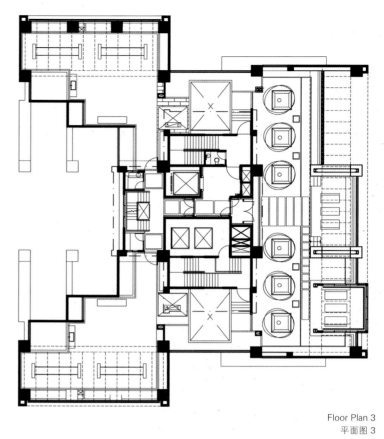

Floor Plan 3
平面图 3

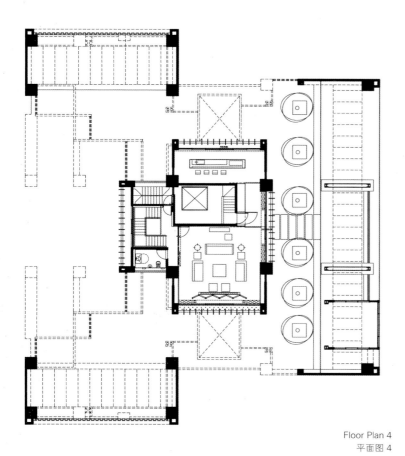
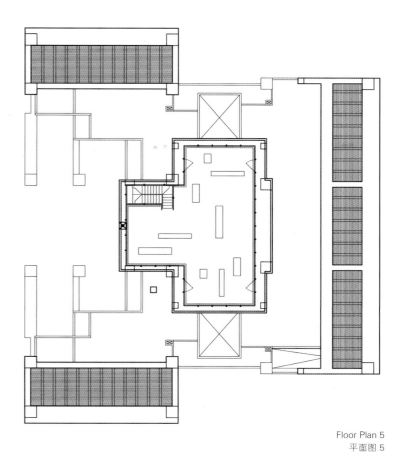

Floor Plan 4
平面图 4

Floor Plan 5
平面图 5